Hurricane
CALCULUS

John B. Hahn   Terry J. Dunlap, Jr.   Steven P. Matyus

Published by Prometheus Enterprises, Inc.
P.O. Box 357
Hinckley, Ohio 44233
Telephone: 1-800-393-3415

Manufactured in the United States of America by R.R. Donnelley & Sons Company

The cover was designed by Mary Jean Hahn

Five percent of the profits generated from the sale of this book will be donated in the form of educational scholarships on an annual basis. Another five percent will be set aside in a non-profit account for the eventual formation of the Prometheus Foundation.

ISBN 1-886783-00-4

Library of Congress Catalog Card Number: 94-80124

The Prometheus Seal, which appears on the back cover, was designed by Michael R. Hahn

ACKNOWLEDGMENTS

We would like to thank the following people for helping us complete this project:

Kenneth G. Hahn, Jr.
Mary Jean Hahn
J.S. Herdman
John Ciprian
Christine Pham Hahn
Sally Seymour

Special thanks to the following individuals and their organizations for their professional advice and support:

Karen Craun
R.R. Donnelley & Sons Company
Harrisonburg, Virginia

Elizabeth A. Dellinger
Benesch, Friedlander, Coplan, & Aronoff
Cleveland, Ohio

Dave Majoy
Citizens Bank Co.
Sandusky, Ohio

This book is dedicated to all those students who never had the opportunity to pursue careers in science or engineering because they failed to make it through first year calculus.

If you have any suggestions, would like to comment on this book, or would like to obtain additional copies, please contact:

Prometheus Enterprises, Inc.
P.O. Box 357
Hinckley, Ohio 44233
U.S.A.

Telephone: 1-800-393-3415

E-mail: REQ91@aol.com

Table of Contents

Preface

The purpose of this book is to help you, the student, learn first year calculus as quickly and easily as possible. An emphasis is placed on explaining all the necessary concepts and problems you need to know in English, providing you with step-by-step problem solving guidelines, as well as showing you plenty of completely worked out examples. In short, this book teaches you calculus in terms you understand.

Hard to believe? Flip through the book. See for yourself. You'll find it's very user-friendly. Read some of the introductions and explanations--they're crystal clear. Check out the story problems in Chapters 3, 4 and 9--they're a bit more laid back than what you're used to. Note the *Real World Applications*--that's right, you can use this stuff out in the real world. And don't forget to look at some of the completely worked out examples--we think you'll find the explanations adjacent to them to be very helpful. By the way, don't skip over Chapter 1 either. You'll get a lot more out of this book and your course if you take fifteen or twenty minutes to review that first chapter.

Good luck, and best wishes!

John B. Hahn
Terry J. Dunlap, Jr.
Steven P. Matyus

January 1995

The Review

In this chapter we provide you with a review of the pre-calculus topics you must understand in order to succeed in calculus. As you will see, all of these topics center around the understanding of functions. "Why?" you ask. Because every calculus problem we encounter in first year calculus involves functions. Simply put: If you don't understand functions, you won't be able to understand calculus.

Study Suggestions:
(1) Begin your study of this chapter by skimming through it twice.
(2) Quickly review the section headings, graphs, and examples.
(3) Read the chapter thoroughly, working through all of the example problems
(4) Review difficult areas.
(5) Put the new information into your long term memory by reviewing the chapter as soon as you finish learning it, then again the next day, a week later, and again a month later.

Imagine jumping out of a plane at 5,000 feet without a parachute. Sound pretty stupid? Well, if you skip over this chapter a very similar academic fate awaits you. Why? Because you won't have the thorough, working knowledge of functions that you need in order to succeed in calculus.

Why do you need such an understanding of functions to succeed? Because functions are encountered in every first year calculus problem. Why are functions so prevalent? Because calculus is the branch of mathematics composed of the study of three mathematical operations-- *limits*, *derivatives*, and *integrals*--which are *used on functions* to determine a variety of things (you'll discover these in the coming chapters). Hence, in order to understand what's going on in calculus, you must first understand functions.

For those of you who agree with us in principle but say you already know all about functions and pre-calculus, we have the following short quiz:

1. What is a function?
2. What is the difference between single and multi-variable functions?
3. What is the difference between an independent and a dependent variable?
4. How do you determine where two lines on a Cartesian coordinate system intersect?

5. What is the formula for the quadratic equation?
6. Name your three favorite trigonometric identities.

How did you do? If you had any difficulties at all, we strongly encourage you take a few moments and work through this introductory chapter with us. Go ahead, put that parachute on. You'll be glad you did when we take off in Chapter 2.

1.1 Review of Algebra: Functions

So what is a function? *A function is a mathematical rule that tells us how two groups of numbers are related to each other.* How? By pointing out that if you take one of the groups of numbers and perform some series of mathematical operations on them, you will obtain the other group. By convention, the group of numbers that gets worked on is referred to as the *domain* of the function, the group that gets produced is referred to as the *range* of the function, and the series of mathematical operations is what we have been calling the mathematical rule-i.e., the function. Let's use a couple of pictures to clarify the concept.

In Fig. 1.1 we have two circles marked x and y with a box in-between them. The circles represent the groups of numbers and the box represents the function. The function takes a number from the domain, performs some

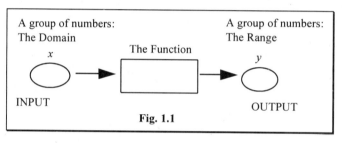

A group of numbers: The Domain

x

The Function

A group of numbers: The Range

y

INPUT

OUTPUT

Fig. 1.1

mathematical operations on it, and then outputs the result-i.e., a number which is an element of the range.

Another way to visualize what's going on here is to think of this whole process in terms of machining. Think of the domain as a bag containing raw material, the function as a machine, and the range as a bag containing finished products, as in Fig. 1.2. You dump raw material into the machine from bag x, the machine works on the raw material and turns it into finished products, which it then spits out into bag y. A simple input/output process. That's the essence of functions.

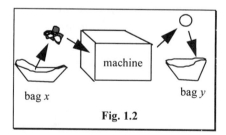

machine

bag x

bag y

Fig. 1.2

Important Details:

1. Inputs and outputs don't vary. When you input a number to a function from its domain, the function outputs a number. *This outputted number is always the same for a given input.* It never varies. For example, if we input the number 2 to a function and the function outputs the number 5, we know that whenever 2 is input, 5 will come out.

Because of this it's sometimes convenient to think of an input and its output as being paired. Hence, we say a number in the domain of a function is *assigned* one, and only one, number in the range of the function. This fact is sometimes referred to as *one-to-one correspondence*. In our example, we would say the number 2 in the domain is assigned the number 3 in the range.

2. Different inputs can produce the same output. It's possible, and fairly common, for two or more numbers in the domain to produce the same number in the range. For example, let's say we input the number 2 to a function and get the number 3 outputted. It's possible to input another number, say -2, and

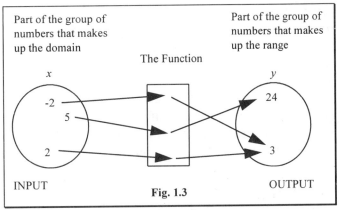

Fig. 1.3

get 3 outputted again. It all depends on the function. Fig. 1.3 above shows part of the domain of a function and its corresponding part of the range to illustrate this point.

3. Variable names. If you use variables to represent numbers in the domain and range, the variable used to represent the domain is called the *independent variable,* and the variable used to represent the range is called the *dependent variable*.

4. Domains. On occasion you'll be asked to determine the domain of a given function. This is not hard as most functions have generic domains like the *real numbers* or the *positive integers*. Sometimes, though, you'll run into functions that have fairly unique domains--example, all the real numbers except 2. Why is this? Because you can't have a number in the domain of a function that results in no output. For example, numbers that are input to a function which result in an output of some number divided by zero or the even root of a negative number are not in the domain of the function since these outputs "do not exist"*. Hence, the way you determine what numbers make up the domain of a function is to:

1. Find all the numbers that cause the function to not exist.
2. List the domain as being all the real numbers except those you found in step 1.

For example, if we have the function $f(x) = 1/x$, zero is not in the domain since $1/0$ does not exist. Every other real number you can think of is fine though, hence the domain of this function is all the real numbers except zero.

---

* You will find different branches of science and other math courses that introduce concepts/notations to account for, and give meaning to, these outputs. For our purposes in this course, however, these types of outputs do not exist.

5. Notation. $y = f(x)$, which is read, "y equals f of x". y is the dependent variable and x is the independent variable. "f" is the name of the function--it's an arbitrary way to identify the function. $f(x)$ is just another term we use to represent numbers in the

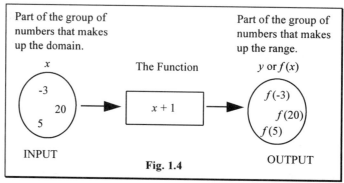

Part of the group of numbers that makes up the domain.

x

The Function

-3
20
5

$x + 1$

Part of the group of numbers that makes up the range.

y or $f(x)$

$f(-3)$
$f(20)$
$f(5)$

INPUT

Fig. 1.4

OUTPUT

range. See Fig. 1.4. Note that if we have two or more functions to work with in the same problem, we will commonly refer to one as "f" and the other as "g". For example, the two functions: $f(x) = x + 1$, and $g(x) = x - 10$.

Examples

1.1

1. $f(x) = x + 1$. $x + 1$ is the "rule". It says you take a number from the domain and add one to it to produce that number's assigned number in the range. For example, 4 is in the domain of this function. To determine what its corresponding number in the range is, we simply run 4 through the function as follows:

$$f(4) = 4 + 1 = 5$$

Observe that the given function is, f(x) = x + 1, and the given input is x = 4. To determine what the function outputs when 4 is input, simply replace every x you see with a 4 and solve.

Hence, when x = 4, f(x) = 5. In other words, when you plug 4 into the function, 5 comes out. Thus, the number 5 in the range is the corresponding number to the number 4 in the domain.

2. Given the function, $f(x) = 2x^3 + 3$, input some values of x (numbers from the domain of the function) and find (produce) their corresponding values of $f(x)$ (numbers from the range of the function).

--when $x = 0$, $f(0) = 2(0)^3 + 3 = 0 + 3 = 3$

--when $x = 5$, $f(5) = 2(5)^3 + 3 = 250 + 3 = 253$

--when $x = -2$, $f(-2) = 2(-2)^3 + 3 = -16 + 3 = -13$

First, note that the domain of this function is all the real numbers--no real number will cause it to not exist-- hence we can just plug in anything. We pick three easy numbers, and as specified above, just replace every x we see with the numbers. The results are the corresponding outputs.

Real World Application: Compound interest is a beautiful thing. For example, did you know that if you found a half-way decent investment that gave a 15% return each year, and you invested $5,000.00 in it when you were 24, by the time you were 65 you would have $1,540,215.39! Pretty cool, eh? So how did we get that number? By taking advantage of the following function:

$V = P(1 + r)^n$
--where V equals your final dollars, P equals your initial investment, r equals your interest rate, and n equals the number of years invested.

For example, if we assume that you're 24 when you have the money to invest and that you'll be 65 when you want to retire, n is constant at 41. If we assume you've found this mutual fund that gives 15% a year, r is constant at 0.15. Hence we have a function for the present situation of:

$V = P(1 + 0.15)^{41} = P(1.15)^{41}$, which can be written: $V = f(P) = P(1.15)^{41}$, (See Fig. 1.5).

Note that P is the independent variable--it represents all of the numbers in the domain of this function. V is the dependent variable, representing all of the numbers in the range. Clearly, there are an infinite number of elements in each of these groups. You're probably only interested in a few combinations, however, and are certainly not interested in seeing all of them. Hence, the function turns out to be a very useful tool for you here. In order to use it, all you have to do is plug in the appropriate dollar value representing your initial investment (a number from the domain), and then sit back and wait for the function to output the corresponding value of the investment at your retirement (a number from the range).

For example, let's say you only have $2,000.00 to invest. To determine how much you'll have at retirement, just plug in: $V = f(P) = f(2000) = 2000(1.15)^{41} = \$616,086.14$

In regards to the above real world application, don't let the change in variable names mess you up. If you compare Fig. 1.5 to Fig. 1.4, you'll see we have the same basic set-up. Why? Because both figures illustrate the mechanics of a function--it just so happens that in Fig 1.5 the

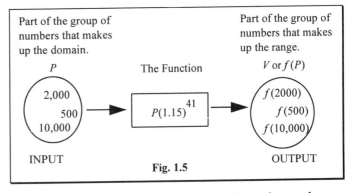

Part of the group of numbers that makes up the domain.

P

2,000
500
10,000

INPUT

The Function

$P(1.15)^{41}$

Part of the group of numbers that makes up the range.

V or $f(P)$

$f(2000)$
$f(500)$
$f(10,000)$

OUTPUT

Fig. 1.5

function being depicting and the variables being used to identify its independent and dependent variable are different from those in Fig. 1.4.

1.2 Review of Algebra: Graphs of Functions

You know what functions are, what some of their basic characteristics are, how they work, as well as how we represent them mathematically. Now we're going to learn how to get a picture of what they look like-i.e., how to graph functions.

So what do we mean, graph a function? We mean plot the numbers from the domain and range of a function on a *Cartesian coordinate system* (Note: There are other types of coordinate systems, but you won't need to worry about them until later in life).

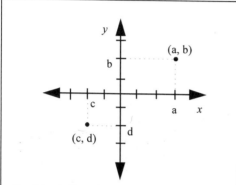

What's a Cartesian coordinate system? You've seen them before, but for the record, a Cartesian coordinate system is a labeled plane. There are two perpendicular axes on the plane, one representing the independent variable (x), the other representing the dependent variable (y). By convention, the independent variable is represented on the horizontal axis while the dependent variable is represented on the vertical axis.

Fig. 1.6: A Cartesian coordinate system. Note that every point in the system can be identified uniquely using ordered-pair notation.

We say the plane is labeled because each point on the plane can be represented uniquely as an ordered pair of numbers--e.g., (a, b), (c, d). (See Fig. 1.6). The first number in the parenthesis is the corresponding x number (coordinate), and the second is the corresponding y number (coordinate) of the point, (x, y).

In order to graph a function on a Cartesian coordinate system, we can take corresponding numbers from the domain and range of the function, form them into ordered-pairs, and plot. For example, the function $f(x) = 2x + 1$ could be graphed using the following ordered pairs from the function:

x	$y = f(x)$	(x, y)
0	1	(0, 1)
1	3	(1, 3)
2	5	(2, 5)
-0.5	0	(-0.5, 0)
-1	-1	(-1, -1)
-2	-3	(-2, -3)
-3	-5	(-3, -5)

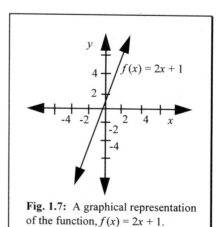

Fig. 1.7: A graphical representation of the function, $f(x) = 2x + 1$.

Some more terminology regarding the Cartesian coordinate system:

1. The point(s) where the graph of the function intersects the *y*-axis is called the *y*-intercept and the point(s) where the graph intersects the *x*-axis is called the *x*-intercept. In Fig. 1.7, the *x*-intercept is the point (-0.5, 0) and the *y*-intercept is the point (0, 1).

2. The point where the *x*- and *y*-axes intersect on the Cartesian coordinate system is called the origin, always the point (0, 0).

3. Sometimes we call the *x*-coordinate of an ordered-pair the *abscissa*, and the *y*-coordinate the *ordinate*.

1.3 Review of Algebra: Information from Graphs of Functions

Pictures are nice, but what sorts of information can we get from graphs of functions? Lots. What follows is a listing of some of the information you can get from graphs of functions that you'll find useful in helping you to solve and/or understand various types of calculus problems.

1. *Slope.* The slope, or steepness, of a straight line is one very important piece of information you can obtain from the graph of a function that is represented by a straight line. As you will remember from algebra I, slope is defined as:

$$m = \frac{y_2 - y_1}{x_2 - x_1}$$

--where (x_1, y_1) and (x_2, y_2) are two separate points on the line.

2. *The equation of a straight line.* The equation of a straight line is:

$$y = mx + b$$

--where *m* is the slope, *b* is the *y*-coordinate of the *y*-intercept (where the line crosses the *y*-axis), and *y* and *x* represent the coordinates of any point on the line (*x, y*).

3. *Distance between two points.* The distance between any two points, (x_1, y_1) and (x_2, y_2), is:

$$\sqrt{(x_1 - x_2)^2 + (y_1 - y_2)^2}$$

4. The point exactly half-way between any two points, (x_1, y_1) and (x_2, y_2), (their *midpoint*), is:

$$\left(\frac{x_1 + x_2}{2}, \frac{y_1 + y_2}{2} \right)$$

5. To determine where two lines intersect, set them equal to each other and solve for the point of intersection. For example, the lines: $f(x) = y = x + 1$ and $g(x) = y = -x + 10$.

Step 1. Set both equations equal to zero:
$$f(x) = y = x + 1$$
$$y - (x + 1) = x + 1 - (x + 1)$$
$$y - x - 1 = 0$$

$$g(x) = y = -x + 10$$
$$y - (-x + 10) = -x + 10 - (-x + 10)$$
$$y + x - 10 = 0$$

The easiest way to set an equation equal to zero is to subtract one side of the equation from both sides. Remember, you can do anything you want to one side of an equation as long as you do it to the other side as well.

Step 2. Since both equations equal zero we can set them equal to each other and solve:

$$y - x - 1 = y + x - 10$$

$$y - x - 1 - (y + x - 10) = 0$$
$$-2x + 9 = 0$$
$$x = 9/2$$

The idea here is that when two equations equal the same thing (in this case zero), their x and y values for that instance are the same. If you think of this in graphical terms, having the same x and y values means having the same point. Hence, in Steps #2 and #3, we're solving the one point that both of these equations have in common--the x and y coordinates of the point of intersection.

Step 3. Plug x back into either of the above equations in Step #1, and solve for y:

$$y - x - 1 = 0 \qquad\qquad y + x - 10 = 0$$
$$y - 9/2 - 1 = 0 \quad \text{or} \quad y + 9/2 - 10 = 0$$
$$y = 11/2 \qquad\qquad\qquad y = 11/2$$

1.4 Review of Algebra: Common Functions & Tricks

In this section we list several common functions you'll run into as we go through calculus. We also mention some tricks we'll use to simplify and solve problems involving them.

1. Common geometric formulas:

Circle of radius r: area $= \pi r^2$, circumference $= 2\pi r$

Sphere of radius r: volume $= \dfrac{4}{3}\pi r^3$, surface area $= 4\pi r^2$

Right circular cylinder of radius r and height h: surface area $= 2\pi r^2 + 2\pi r h$

volume $= \pi r^2 h$

Triangle of base b and height h: area $= \dfrac{1}{2}bh$

2. Logarithms:

$$If\ b^x = y,\ then\ x = \log_b y \qquad \log\left(\frac{1}{a}\right) = -\log a$$

$$\log(ab) = \log a + \log b \qquad \log a^n = n \log a$$

$$\log\left(\frac{a}{b}\right) = \log a - \log b \qquad \log 1 = 0$$

3. *Natural logarithms*: $e = 2.71828$, $e^x = y$, $x = \log_e y = \ln y$

4. *Factoring*: Factoring involves simplifying fractions by representing complex looking expressions in either their numerator or denominator as multiples of two or more simpler expressions, and then canceling. You'll be inundated by examples of this in the next chapter so we won't get into any detail here. Factoring is useful in simplifying functions so they're easier to work with.

5. *The quadratic equation*: The quadratic equation is used to solve the following types of equations as follows:

$$If\ ax^2 + bx + c = 0,\ then\ x = \frac{-b \pm \sqrt{b^2 - 4ac}}{2a}$$

1.5 Review of Trigonometry: Functions in Trigonometry

So far in this chapter we've reviewed the basics of functions, which we all learned about in algebra. Now we shift into a slightly higher gear and review a special category of functions--trigonometric functions--which we all learned about in trigonometry. Lest you panic as memories of past experiences with this topic begin creeping into your consciousness, let us assure you that there's not that much to learn here. Contrary to popular belief, you don't have to be a trigonometry wizard to succeed in calculus, only mildly competent at it. Here's what you need to know:

Trigonometry is based on the idea of getting information out of *angular measurements*. Thus, in order to understand trigonometry, you must first understand angular measurements.

So, what do we mean by angular measurements? Well, if we take a standard Cartesian coordinate system, place a line segment on it that originates at the origin running along the positive *x*-axis, and rotate the segment counter-clockwise to a certain

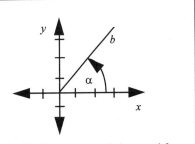

Fig. 1.8: The line segment, *b*, is rotated from its initial position along the *x*-axis to some final position between the *x* and *y* axes. A measure of how far the segment rotated is the angle, α.

position, a measure of how far we've rotated the segment is the angle the segment now forms with respect to its original position, as is shown in Fig. 1.8 above.

How do we quantify angles? The two ways we'll see in calculus are via *degree* and *radian* measures. As you will recall, there are 360 degrees in a circle. What do we mean by this? We mean that if you were to divide a circle into 360 equal parts about its circumference, every unit you traveled (every 1/360 you went around the circle) would be equal to one degree of rotation. Nothing complicated or new about this.

Radians, which you may or may not have any recollection of, are defined a bit differently: If you move along the circumference of a circle a distance equal to the circle's radius, your rotation about the circle is equal to one radian. See Fig. 1.9.

As you'll probably recall, there are 2π radians in every circle. Where do we get this? For those of you who are interested, we supply the following derivation:

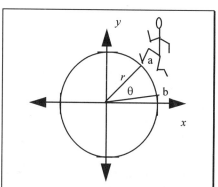

Fig. 1.9: You start running around a circle of radius r, from a point b on the circle. When you get to a point a on the circle, you've rotated 1 radian about the circle if the distance between a and b (in terms of arc length) is equal to r.

1. $\pi = \dfrac{\text{circumference of circle}}{\text{circle's diameter}}$

   $\therefore$ circumference of circle $= 2\pi r$

2. arc length $= s = r\theta$

   --where θ is the angle of rotation.

3. If you rotate all the way around a circle, the arc length equals the circumference. Hence:

   $r\theta = 2\pi r$

   $\theta = \dfrac{2\pi r}{r} = 2\pi$

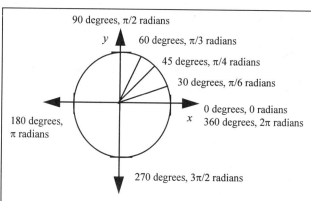

Fig. 1.10: This picture shows the correspondence between degrees and radians as you rotate counter-clockwise about a circle.

To sum up then, there are 360 degrees in a circle, and 2π radians. To get a look at how various degree and radian measurements compare, see Fig. 1.10 above. To convert between degrees and radians, we do the following:

$$\text{radians} = \text{degrees} \cdot \frac{\pi}{180}; \quad \text{and degrees} = \text{radians} \cdot \frac{180}{\pi}$$

Now that we're all up to speed on angular measurements we can begin our review of the functions that are used in conjunction with them--trigonometric functions. There are six basic trigonometric functions which you need to be aware of. These are defined as follows.

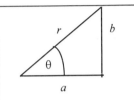

Fig. 1.11: The trigonometric functions are defined using a triagle with sides of length a, b, and r, and angle θ.

$$\sin\theta = \frac{b}{r} \qquad \csc\theta = \frac{r}{b} \quad (b \neq 0)$$

$$\cos\theta = \frac{a}{r} \qquad \sec\theta = \frac{r}{a} \quad (a \neq 0)$$

$$\tan\theta = \frac{b}{a} \quad (a \neq 0) \qquad \cot\theta = \frac{a}{b} \quad (b \neq 0)$$

Trigonometric Identities

One of the neat things about trigonometric functions is that they are related to each other in many ways, defined by trigonometric identities. The more basic trigonometric identities which you should be aware of are listed below. Memorization of these identities is not required. Just make sure you know about them so you can access the information when you need it in future chapters.

Fundamental Identities:

$$\sin^2\theta + \cos^2\theta = 1 \qquad \sec^2\theta - \tan^2\theta = 1 \qquad \csc^2\theta - \cot^2\theta = 1$$

$$\tan\theta = \frac{\sin\theta}{\cos\theta} \qquad \cot\theta = \frac{\cos\theta}{\sin\theta} \qquad \csc\theta = \frac{1}{\sin\theta}$$

$$\sec\theta = \frac{1}{\cos\theta} \qquad \cot\theta = \frac{1}{\tan\theta}$$

Useful Theorems:

$$\sin 2\theta = 2\sin\theta \cos\theta$$

$$\cos 2\theta = \cos^2\theta - \sin^2\theta = 1 - 2\sin^2\theta = 2\cos^2\theta - 1$$

$$\tan 2\theta = \frac{2\tan\theta}{1 - \tan^2\theta}$$

$$\sin(\theta \pm \partial) = \sin\theta \cos\partial \pm \cos\theta \sin\partial$$

$$\cos(\theta \pm \partial) = \cos\theta \cos\partial \mp \sin\theta \sin\partial$$

Bonus Info: What we've been showing you in this chapter are single-variable functions. In other words, functions that have only one independent variable. Multi-variable functions are functions that have two or more independent variables. For example, the function, $y = f(x, z) = x + 2z - 1$, is an example of a multi-variable function. y is the dependent variable, x and z are the independent variables. In first year calculus we deal with single variable functions. In third semester calculus, we apply everything we've learned in first year calculus to multi-variable functions.

Limits

Having successfully completed our review of functions, we begin our study of calculus by introducing limits, the first of the three mathematical operations that comprise calculus. As you will see, limits are the most theoretical of the three operations--a kind way of saying they're essentially useless in the real world. Nevertheless, we find ourselves compelled to study them much as we do spelling when we're learning how to write. You see, just as letters are used to make up words, limits are used to make up the other two mathematical operations that comprise calculus. By understanding limits, we hopefully gain a better appreciation for the other two. In other words, we have a mathematical spelling lesson to learn. It's not difficult. Relax and enjoy.

To give you a flavor for what lies ahead, we're going to begin our study of limits with an example from another academic topic, ancient history, that brings to life one of the key concepts of this chapter. Here it is: In the 5th Century B.C., in the little Italian town of Elea, there lived a philosopher who went by the name, Zeno of Elea. Now Zeno was not your average philosopher. Besides being a bit on the daring side--he once joined a conspiracy to oust the tyrant Nearchus from Elea (it failed, Zeno was tortured)--Zeno had a knack for making obvious observations that were, in reality, not so obvious. For example, he once pointed out that *if a traveler goes halfway to his destination each day, he will never reach his destination since there is always another halfway to go.*

Trivial? Maybe in the context of life. But in the context of upper level mathematics (read: calculus) it is a statement of profound insight. Why? Because it explains the mechanism behind limits, the first of the three mathematical operations that comprise calculus.

2.1 Limits--The Basic Idea

Before we expand on Zeno's profound insight, we need to answer the question you're all undoubtedly pondering: What is a limit?

A limit is a mathematical operation used to determine what value, if any, a function will output when numbers infinitely close to some specified number are input to the function.

In other words, if we're given a function, say $f(x) = x^2$, and we want to know what number the function will output when numbers infinitely close to a specified number, say $x = 2$, are input to the function, we can use a limit to figure this out. Fig. 2.1 illustrates the problem.

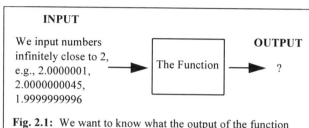

Fig. 2.1: We want to know what the output of the function will be when numbers infinitely close to the number 2 are input to the function.

So now you're asking, when would this ever be important? We give you the following example:

Real World Application: Tickling a sleeping dragon's tail. Sounds kind of dangerous doesn't it. Well, in late 1944 at Los Alamos an experiment thought up by Otto Frisch became known as the Dragon experiment because it struck fellow scientist Richard Feyman as being very similar to tickling a sleeping dragon's tail. Here's the story:

In the early 1940's the United States had assembled the finest scientific minds in the country at Los Alamos New Mexico for the express purpose of developing the world's first atomic bomb. In late 1944 the group was faced with the problem of determining how much Uranium was required to obtain critical mass-i.e., essentially, the minimum amount of fissionable material required to induce an atomic explosion. One of the scientists, Otto Frisch, came up with a very creative solution.

Otto noted that if one was to take a hunk of Uranium and form it into a ring, then drop a ball of Uranium through the ring, during the split second when the ball was passing through the ring, more Uranium would be together and the subsequent reaction could be monitored. By dropping progressively greater amounts of Uranium through the ring, critical mass could be approached. Thus, the scientists would be able to accurately determine the amount of Uranium required for the bomb. Of course, great care had to be taken never to reach critical mass, as that would have resulted in a most untimely large bang.[1]

See the connection? In the Dragon Experiment it was quite desirable to know the output of the given apparatus when objects very close to a specified object, critical mass, were input. It was also important that the specified object never be input. Strikingly similar to wanting to know the output of a function when numbers infinitely close to a specified number are input to the function, yes?

---

[1] Rhodes, Richard. *The Making of the Atomic Bomb*, Simon & Schuster, 1986. pp. 610-611.

Next question: How does a limit work? This is best seen by example, so suppose we're given the function $f(x) = x^2$, and we want to know what the value of the function will be when numbers infinitely close to $x = 2$ are input to the function. Here's what the limit does:

The limit arbitrarily picks some number on one side of the specified number and inputs that number into the function. In this case, say the number 6.

Fig. 2.2: A number line depicting where the specified number and the arbitrary input are in regards to each other.

It then picks another number to input, this time, though, with the constraint that the number be exactly halfway between the previous number and the specified number, in this case the number 4.

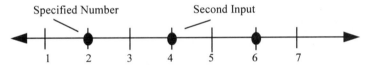

Fig. 2.3: The second number input to the function is always exactly halfway between the arbitrary input and the specified number.

The function repeats this step an infinite number of times--each time inputting to the function a number exactly halfway between the pervious input and the specified number. In this way, the specified number is never input to the function because it is never reached (similar to the traveler who goes halfway to his destination each day, never getting there, yes?) We have a term for this phenomena--*approaching*. We say the inputted values *approach* the specified number.

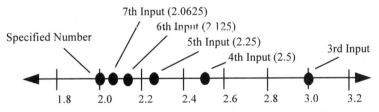

Fig. 2.4: From the second input on, the limit inputs the number exactly halfway between the previous input and the specified number. In this figure we show the first six inputs (resolution prevents us from showing any more). The point is that the specified number is never input to the function, but the numbers infinitely close to it are. The term for this method of inputting numbers is *approaching*.

Once the limit gets done approaching the specified number from one side, it then approaches it from the other side.

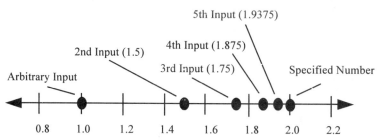

Fig. 2.5: Approaching the specified number from the other side, from a lesser number. The same rules apply as when approaching from a greater number--after the arbitrary input, each succeeding input is exactly halfway between the previous input and the specified number. As in Fig. 2.4, resolution prevents us from showing any more than the first few inputs.

As you can imagine, inputting the two infinite sets of numbers into the function results in two infinite sets of outputs from the function. If both of these sets of outputs move towards the same number, then this number is said to be the limit of the function for the specified number.

Note: In some cases the two sets of outputs do not move towards the same number. In these cases there is said to be no limit of the function for the specified number.

So, in our above case with the function $f(x) = x^2$, and $x = 2$ as our specified input, the limit would work as follows:

Limit mechanism:

Approaching from the greater value	Resulting output from the function	Approaching from the lesser value	Resulting output from the function
Input: x	Output: $f(x)$	Input: x	Output: $f(x)$
4	16	0	0
3	9	1	1
2.5	6.25	1.5	2.25
2.25	5.0625	1.75	3.0625
2.125	4.515625	1.875	3.515625
2.0625	4.2539063	1.9375	3.7539063
2.03125	4.1259766	1.96875	3.8759766
2.015625	4.0627441	1.984375	3.9377441
2.0078125	4.0313110	1.9921875	3.9688110
2.00390625	4.0156401	1.9960938	3.9843903

Table 2.1: *Remember, the limit approaches the specified input, in this case 2, from a greater number, and records the results. It then approaches the specified number from a lesser number and records the results. If both sets of results move toward the same number, that number is said to be the limit of the function for the specified input. In this case, 4 is the limit.*

In this example, both sets of outputs from the function move towards the same number, 4, as the specified number, 2, is approached from both sides. Hence, 4 is the limit of the function for this specified input. Formally, we say that, "the limit of the function $f(x) = x^2$, as x approaches 2, is 4".

Notation: $\qquad \lim\limits_{x \to a} f(x) = b$

--which is read, the limit of the function f of x, as x approaches a, equals b.

After going through this introduction on limits, you should have a good idea of what limits are and how they work. Now the obvious question of how we go about using them to solve problems arises. It will thus be the purpose of the next several sections to help you learn how to do this. We begin with the simplest approach (IT'S EASY!!):

To solve limit problems, plug the specified number into the function.

1. If you get a number, that is the limit of the function.

2. If you get a constant divided by zero, the function has no limit for the specified number.

3. If you get zero divided by zero, factor the top and bottom, cancel, and plug in again repeating the process. If you can't factor, try another algebra trick like the conjugate method (which is illustrated in the examples below). If you still can't factor after all that, then resort to graphing the function using the limit mechanism technique as shown above.

Examples   **2.1**

1. $\lim\limits_{x \to 1} \dfrac{1}{2}x^2 = ?$

$\lim\limits_{x \to 1} \dfrac{1}{2}x^2 = \dfrac{1}{2}(1)^2 = \dfrac{1}{2}$

In this question we're asked to find the limit of the function 1/2 x squared as x approaches 1. Using the above guidelines, we plug 1 into the function and solve. In so doing, we get the number 1/2. Thus, 1/2 is the limit of this function when x approaches 1.

2. $\lim\limits_{x \to 3} 5 = ?$

$\lim\limits_{x \to 3} 5 = 5$

Trick question. We're asked to find the limit of the function 5 as x approaches 3. There's no variable, no x, in the function though, so there's no place to plug in the 3. Hence, the limit is just the constant, 5.

3. $\lim\limits_{x \to 0} 7x^3 = ?$

$\lim\limits_{x \to 0} 7x^3 = 7(0)^3 = 0$

Simple example just to drive home the point that the first thing you do with limit problems is plug in the specified number to see if you get a number. If you do, that's the limit.

4. $\lim\limits_{x\to 1} \dfrac{x^2-1}{x-1} = ?$

$\lim\limits_{x\to 1} \dfrac{x^2-1}{x-1} = \dfrac{(1)^2-1}{1-1} = \dfrac{0}{0}$

We begin solving this problem, as in the above examples, by plugging in the specified number. When we do this, however, we get zero divided by zero, indicating that we must factor.

$\lim\limits_{x\to 1} \dfrac{x^2-1}{x-1} = \lim\limits_{x\to 1} \dfrac{(x-1)(x+1)}{(x-1)}$

The idea of factoring is to simplify the fraction by finding common terms in both the numerator and denominator so that they can be canceled out. Here, the common term is (x - 1).

$= \lim\limits_{x\to 1} (x+1)$

$= (1)+1 = 2$

Once the factoring is complete, we plug in the specified number again, this time obtaining our answer.

5. $\lim\limits_{x\to 0} \dfrac{1}{x} = ?$

$\lim\limits_{x\to 0} \dfrac{1}{x} = \dfrac{1}{0} = \text{DNE}$

Like usual, we plug in the specified number, but this time we get one divided by zero. As indicated by the above guidelines, if you obtain a constant divided by zero, the function has no limit. Hence, we write as our answer, DNE, which stands for "Does Not Exist".

6. $\lim\limits_{x\to 0} \dfrac{|x|}{x} = ?$

$\lim\limits_{x\to 0} \dfrac{|x|}{x} = \dfrac{|0|}{0}$

In this problem when we plug in the specified number we get zero divided by zero. However, when we try to factor, we can't. Thus, we must graph the function to solve this problem.

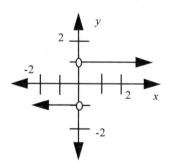

The easiest way to graph this function is to plug in various x values, obtain their corresponding y values, and plot. When we do this we find that the graph of the function is discontinuous at x equals zero, our specified number. Further, we see that as x equals zero is approached from both sides, both outputs do not move towards the same number--one moves towards -1, the other 1. Thus, we say that the limit of this function Does Not Exist.

Having gotten your feet wet on this initial barrage of examples, you're probably thinking that this isn't so bad. And guess what, you're right! As we said earlier, this stuff isn't that hard. Now, it's true that these initial examples were not the most difficult, however, if you understand them the rest of the limit problems will come naturally for you. To help you through the rest of the limit problems, the following list of theorems is provided. Look them over, but don't try to memorize them. Just go through the rest of the examples to see these theorems in action.

Useful Theorems

a. The limit of a constant is the constant.

b. Let $\lim\limits_{x \to a} f(x) = A$, let $\lim\limits_{x \to a} g(x) = B$, let c = constant

  i. $\lim\limits_{x \to a}\left[f(x) + g(x)\right] = A + B$

  ii. $\lim\limits_{x \to a}\left[f(x) - g(x)\right] = A - B$

  iii. $\lim\limits_{x \to a}\left[f(x) \cdot g(x)\right] = A \cdot B$

  iv. $\lim\limits_{x \to a}\left[f(x) \div g(x)\right] = A \div B,\ B \neq 0$

  v. $\lim\limits_{x \to a}\left[c \cdot f(x)\right] = c \cdot A$

c. If n is a positive integer and $\lim\limits_{x \to a} f(x) > 0$, then

$$\lim\limits_{x \to a}\left[f(x)\right]^n = \left[\lim\limits_{x \to a} f(x)\right]^n$$

d. If n is an ODD positive integer, then $\lim\limits_{x \to a}\left[f(x)\right]^{\frac{1}{n}} = \left[\lim\limits_{x \to a} f(x)\right]^{\frac{1}{n}}$

e. If n is an EVEN positive integer AND $f(a) > 0$, then

$$\lim\limits_{x \to a}\left[f(x)\right]^{\frac{1}{n}} = \left[\lim\limits_{x \to a} f(x)\right]^{\frac{1}{n}}$$

f. If n is an EVEN positive integer AND $f(a) < 0$, then

$$\lim\limits_{x \to a}\left[f(x)\right]^{\frac{1}{n}} = \text{Does Not Exist}$$

Examples   **2.1, continued...**

7. $\lim\limits_{x \to 2}\left(2x^2 + 5x\right) = ?$

$\lim\limits_{x \to 2}\left(2x^2 + 5x\right) = 2(2)^2 + 5(2)$

$\qquad = 8 + 10 = 18$

In this problem we have two functions, $2x^2$ and $5x$, being added together. Taking advantage of the limit addition theorem above, this problem is readily solved. We simply find the limit of each function and then add them together. In this case, plugging in the specified number does the trick.

8. $\lim\limits_{x \to 2}\left(7x^2 - x\right) = ?$

$\lim\limits_{x \to 2}\left(7x^2 - x\right) = 7(2)^2 - (2)$

$\qquad = 28 - 2 = 26$

Again in this problem we have two functions, $7x^2$ and x, this time being subtracted. Taking advantage of the limit subtraction theorem above, we simply find the limit of each function and then subtract them.

9. $\lim\limits_{x \to 1} \left(3x^2 + \left(\dfrac{x^2 - 1}{x - 1} \right) \right) = ?$

$\lim\limits_{x \to 1} \left(3x^2 + \left(\dfrac{x^2 - 1}{x - 1} \right) \right) = 3(1)^2 + \left(\dfrac{(1)^2 - 1}{1 - 1} \right) = 3 + \dfrac{0}{0}$

This is a slightly more complicated problem. We have two functions being added together, so to solve we need to find the limit of each and then add them together. Nothing new there.

$\lim\limits_{x \to 1} \left(\dfrac{x^2 - 1}{x - 1} \right) = \lim\limits_{x \to 1} \dfrac{(x - 1)(x + 1)}{(x - 1)} = \lim\limits_{x \to 1} (x + 1) = 1 + 1 = 2$

$\therefore \ \lim\limits_{x \to 1} \left(3x^2 + \left(\dfrac{x^2 - 1}{x - 1} \right) \right) = 3 + 2 = 5$

As you can see, the difficulty arises in that the second function's limit is not found simply by plugging in the specified number. An additional step--factoring--is required to find the second function's limit before the two limits can be added together.

10. $\lim\limits_{x \to 3} x^2 \left(x^3 - 2x^2 + 5x - 1 \right) = ?$

$\lim\limits_{x \to 3} x^2 \left(x^3 - 2x^2 + 5x - 1 \right) = (3)^2 \left[\left(3^3 - 2(3)^2 + 5(3) - 1 \right) \right]$

$= 9[27 - 18 + 15 - 1] = 207$

Here we have a limit problem involving multiplication, addition, and subtraction of functions. Since all of the limits of these functions can be found by simply plugging in the specified number, we solve for them and take advantage of the above theorems at the same time.

11. $\lim\limits_{x \to 1} (x + 3)(x - 5) = ?$

$\lim\limits_{x \to 1} (x + 3)(x - 5) = (1 + 3)(1 - 5)$

$= -16$

Relatively simple limit problem involving multiplication of two functions. Again, both limits can be found by plugging in the specified number, so we do that and take advantage of the above theorem.

12. $\lim\limits_{x \to 2} \dfrac{x^2}{x - 1} = ?$

$\lim\limits_{x \to 2} \dfrac{x^2}{x - 1} = \dfrac{(2)^2}{2 - 1} = \dfrac{4}{1} = 4$

Our first problem involving division. Nothing complicated. The limits of both functions are found by simply plugging in the specified number, so we do this while we take advantage of the limit division theorem above.

13. $\lim\limits_{x \to 1} \left(2x^4 - 3 \right)^{\frac{1}{3}} = ?$

$\lim\limits_{x \to 1} \left(2x^4 - 3 \right)^{\frac{1}{3}} = \left(2(1)^4 - 3 \right)^{\frac{1}{3}}$

$= (-1)^{\frac{1}{3}} = -1$

Our first problem involving functions raised to powers less than 1. In this case, the denominator of the fraction to which this function is being raised is an ODD positive integer (the number 3). Hence, according to the above theorems, all we have to do is solve for the limit of the function, then raise that answer to the given power (in this case, 1/3).

14. $\lim\limits_{x \to 2} (x-5)^{\frac{1}{4}} = ?$

If $\lim\limits_{x \to 2} (x-5) > 0$, then $\lim\limits_{x \to 2} (x-5)^{\frac{1}{4}} = \left[\lim\limits_{x \to 2} (x-5)\right]^{\frac{1}{4}}$

Our second problem involving functions raised to powers less than one has an even numbered denominator in the fraction to which the function is being raised.

$\lim\limits_{x \to 2} (x-5) = (2-5) = -3$

∴ The limit Does Not Exist

Hence, according to the above theorems, we must check to see if the function is greater than 0 when the specified number is input. In this case it is not. Thus, the limit does not exist.

15. $\lim\limits_{x \to 3} (2x-1)^{\frac{1}{2}} = ?$

If $\lim\limits_{x \to 3} (2x-1) > 0$, then $\lim\limits_{x \to 3} (2x-1)^{\frac{1}{2}} = \left[\lim\limits_{x \to 3} (2x-1)\right]^{\frac{1}{2}}$

Our third problem involving functions raised to powers less than one again has an even number in the denominator of the fraction to which the function is being raised.

$\lim\limits_{x \to 3} (2x-1) = (2(3)-1) = 5$

∴ $\lim\limits_{x \to 3} (2x-1)^{\frac{1}{2}} = (5)^{\frac{1}{2}}$

As in the last example, we first check to see if the function is greater than 0 when the specified number is input. It is. Hence, all we have to do is raise this result to the given power to obtain our answer.

16. $\lim\limits_{x \to 0} \dfrac{\dfrac{1}{\sqrt{4-x}} - \dfrac{1}{2}}{x} = ?$

We begin solving this problem as we normally do, plugging in the number that is being approached. In so doing, we get zero divided by zero, indicating we must factor. However, upon examination of the original function, we note that standard factoring is not possible. We could graph, as we did in a previous example, but there is an easier way. An algebra trick that will enable us to factor--the conjugate technique.

$\lim\limits_{x \to 0} \dfrac{\dfrac{1}{\sqrt{4-x}} - \dfrac{1}{2}}{x} = \dfrac{\dfrac{1}{\sqrt{4-0}} - \dfrac{1}{2}}{0} = \dfrac{\dfrac{1}{2} - \dfrac{1}{2}}{0} = \dfrac{0}{0} = \dfrac{0}{0}$

The conjugate technique relies on the fact that you can multiply the numerator of a fraction by anything without changing the value of the fraction, as long as you do the same with the denominator. For example, the fraction 3/4 can be multiplied by 5/5, 3/3, (10x-2)/(10x-2), without changing the value of the fraction. Why, because 5/5, 3/3, and (10x-2)/(10x-2) are all equal to 1, and anything multiplied by 1 equals itself.

Hurricane Calculus

In the conjugate technique, you find the conjugate of the "problem area", and multiply the top and bottom of the fraction by it. For example, the conjugate of (x - 2) is (x + 2). The conjugate of $(5x^3 + 7x^5)$ is $(5x^3 - 7x^5)$. In this particular example, the conjugate is thus:

$$\frac{1}{\sqrt{4-x}} + \frac{1}{2}$$

To solve this problem then, we simply multiply the top and bottom of the fraction by the conjugate, then rearrange terms until we can cancel a variable.

$$\lim_{x \to 0} \left[\frac{\frac{1}{\sqrt{4-x}} - \frac{1}{2}}{x} \times \frac{\frac{1}{\sqrt{4-x}} + \frac{1}{2}}{\frac{1}{\sqrt{4-x}} + \frac{1}{2}} \right] = \lim_{x \to 0} \frac{\frac{1}{4-x} - \frac{1}{4}}{x \left(\frac{1}{\sqrt{4-x}} + \frac{1}{2} \right)}$$

We begin by multiplying the top and bottom of the fraction by the conjugate of the numerator. Then, we rearrange the terms in the numerator so we can cancel something in the denominator.

$$= \lim_{x \to 0} \frac{\frac{4-(4-x)}{4(4-x)}}{x \left(\frac{1}{\sqrt{4-x}} + \frac{1}{2} \right)}$$

$$= \lim_{x \to 0} \frac{\frac{x}{4(4-x)}}{x \left(\frac{1}{\sqrt{4-x}} + \frac{1}{2} \right)}$$

We note that after much manipulation, it is possible to cancel an x in the numerator with the x in the denominator.

$$= \lim_{x \to 0} \frac{\frac{1}{4(4-x)}}{\left(\frac{1}{\sqrt{4-x}} + \frac{1}{2} \right)}$$

After all the multiplying, rearranging, and the canceling, we have the same function, in a different, solvable form.

$$\lim_{x \to 0} \frac{\frac{1}{4(4-x)}}{\left(\frac{1}{\sqrt{4-x}} + \frac{1}{2} \right)} = \frac{\frac{1}{4(4-0)}}{\left(\frac{1}{\sqrt{4-0}} + \frac{1}{2} \right)} = \frac{\frac{1}{16}}{\frac{1}{2} + \frac{1}{2}} = \frac{1}{16}$$

Hence, we begin again, plugging in 0 to solve.

Additional Problems   2.1

Find the limits in the following problems if they exist.

1. $\lim\limits_{x \to 0} 3$

2. $\lim\limits_{x \to -1} 5$

3. $\lim\limits_{x \to 24} 1020$

4. $\lim\limits_{x \to -19} \dfrac{5}{17}$

5. $\lim\limits_{x \to 3} \pi$

6. $\lim\limits_{x \to 2} x$

7. $\lim\limits_{x \to -3} (-2x)$

8. $\lim\limits_{x \to 2} (12 + x)$

9. $\lim\limits_{x \to 4} (x - 7)$

10. $\lim\limits_{x \to 0} \left(3x + \dfrac{1}{2}\right)$

11. $\lim\limits_{x \to 4} (725)(x - 4)$

12. $\lim\limits_{x \to 2} (x^3 + 2x - 1)$

13. $\lim\limits_{x \to 6} (3x^2 - x - 5)$

14. $\lim\limits_{x \to 3} (x^2 + 5x)$

15. $\lim\limits_{x \to 0} \dfrac{(x + 9)^2}{17x}$

16. $\lim\limits_{x \to -3} \dfrac{(x^2 - 9)}{(x + 3)}$

17. $\lim\limits_{x \to 4} \dfrac{(x^2 - 16)}{(x - 4)}$

18. $\lim\limits_{x \to 5} \dfrac{(x^2 - 25)}{(x - 5)}$

19. $\lim\limits_{x \to 3} \dfrac{x^2}{24x + 9}$

20. $\lim\limits_{x \to 2} \dfrac{(x - 2)}{(x^3 - 8)}$

21. $\lim\limits_{x \to 3} \dfrac{(x - 3)}{(x^3 - 27)}$

22. $\lim\limits_{x \to 5} \dfrac{(x - 5)}{(x^3 - 125)}$

23. $\lim\limits_{x \to -2} \dfrac{(x + 2)}{(x^3 + 8)}$

24. $\lim\limits_{x \to -3} \dfrac{(x + 3)}{(x^3 + 27)}$

25. $\lim\limits_{x \to 0} \dfrac{\left(4x^8 - 7x^4 - x^2 + 19\right)}{\sqrt{15x^3 - x^2 + 9x + 3}}$

26. $\lim\limits_{x \to -2} \dfrac{\left(4x^3 + 32\right)}{\left(4x^4 - 64\right)}$

27. $\lim\limits_{x \to -3} \dfrac{\left(x^3 + 27\right)}{\left(x^4 - 81\right)}$

28. $\lim\limits_{x \to 2} \dfrac{\left(x^3 - 2x^2 + 4x - 8\right)}{(x - 2)}$

29. $\lim\limits_{x \to 2} \dfrac{\left(3x^2 - 10x + 8\right)}{\left(x^3 - 2x^2 - 2x + 4\right)}$

30. $\lim\limits_{x \to 5} \dfrac{5 - x}{\left(x^4 - x^{\frac{1}{8}} + 67\right)}$

31. $\lim\limits_{x \to 5} \dfrac{1}{x - 5}$

32. $\lim\limits_{x \to 0} \dfrac{\left(\dfrac{1}{\sqrt{3 + x}} - \dfrac{1}{\sqrt{3}}\right)}{x}$

33. $\lim\limits_{x \to 0} x^{12}$

34. $\lim\limits_{x \to 1} (x - 1)^{\frac{1}{2}}$

35. $\lim\limits_{x \to 1} (x - 2)^{\frac{1}{2}}$

36. $\lim\limits_{x \to 5} (2x - 11)^{\frac{1}{3}}$

41. $\lim\limits_{x \to 0} \dfrac{|x - 1|}{x - 1}$

45. $\lim\limits_{x \to 0} \dfrac{4 - \sqrt{16 + x}}{x}$

37. $\lim\limits_{x \to 0} (x - 5)^{\frac{1}{9}}$

42. $\lim\limits_{x \to -1} \dfrac{(x + 1)^3}{x^3 + 1}$

46. $\lim\limits_{x \to 1} \dfrac{-2 + \sqrt{x + 3}}{x - 1}$

38. $\lim\limits_{x \to 1} (2x^2 - 17)^{\frac{1}{4}}$

43. $\lim\limits_{x \to 1} \dfrac{x^3 - 1}{(x - 1)^{\frac{1}{2}}}$

47. $\lim\limits_{x \to 2} \left(\dfrac{1}{x - 2} - \dfrac{4}{x^2 - 4} \right)$

39. $\lim\limits_{x \to 2} (x^2)^3$

44. $\lim\limits_{x \to 1} \dfrac{x^3 - 1}{x^2 - 1}$

40. $\lim\limits_{x \to 1} \dfrac{|x - 1|}{x - 1}$

2.2 One-Sided Limits

Now that you know what limits are, we expand our study to a special kind of limit: *one-sided limits*. Don't be intimidated by these complex sounding creatures. Outside of their name, there's nothing complex about them at all. In fact, we've already seen them, you just weren't told. You see, *one-sided limits are simply limits who have inputs that approach the specified number from one side, as opposed to both sides like a regular limit.* Those with inputs approaching the specified number from the left (less than the specified number) are called *left hand limits,* while those with inputs approaching from the right (greater than the specified number) are called *right hand limits.*

Notation: Right Hand Limit: $\lim\limits_{x \to a^+} f(x) = b$

Left Hand Limit: $\lim\limits_{x \to a^-} f(x) = b$

--which is read: The limit of f of x, as x approaches a from the right (left) equals b, where a and b are constants.

Pay close attention to the + and - signs used to differentiate between the two types of limits. Remember, the + sign indicates a right hand limit while a - sign indicates a left hand limit.

Mechanism of the One-sided Limits

Given: $y = f(x) = x^2$

Find: $\lim\limits_{x \to 2^+} x^2$, and $\lim\limits_{x \to 2^-} x^2$

Solution:
From the table we see that the right hand limit equals 4, and the left hand limit equals 4.

Approaching from the right:		*Approaching from the left:*	
Input: x	Output: $f(x)$	Input: x	Output: $f(x)$
4	16	0	0
3	9	1	1
2.5	6.25	1.5	2.25
2.25	5.0625	1.75	3.0625
2.125	4.515625	1.875	3.515625
2.0625	4.2539063	1.9375	3.7539063
2.03125	4.1259766	1.96875	3.8759766
2.015625	4.0627441	1.984375	3.9377441
2.0078125	4.0313110	1.9921875	3.9688110
2.00390625	4.0156401	1.9960938	3.9843903

Some important details about one-sided limits: **For a function to have a limit, both the left AND right hand limits for the desired input must exist and give the SAME prediction**. If they don't, then the function is *discontinuous* at that point, and there is no limit. In the example above, a limit exists since both the left and right hand limits exist, and since they give the same prediction.

You see, nothing complicated here. One-sided limits just give you another way of looking at limit problems. The methods of solving regular limits apply here as well with only one slight complication: *In the case of problems in which variables are raised to powers less than one, you have to mentally picture what is going on and cannot simply plug in the number that the limit is approaching.* The examples below illustrate how you go about dealing with this.

Examples **2.2**

1. $\lim\limits_{x \to 4^-} (x-4)^{\frac{1}{2}}$

Since x is approaching 4 from the left, all of the inputs are less than 4. Therefore, when 4 is subtracted from them in the problem, a negative number results. Since the square root of a negative number is imaginary, the limit here does not exist.

$\lim\limits_{x \to 4^-} (x-4)^{\frac{1}{2}} = \text{DNE}$

2. $\lim\limits_{x \to 4^+} (x-4)^{\frac{1}{2}}$

Since x is approaching 4 from the right, all of the inputs are greater than 4. Therefore, when 4 is subtracted from them in the problem, a positive number results and the square root can be found.

$\lim\limits_{x \to 4^+} (x-4)^{\frac{1}{2}} = (4-4)^{\frac{1}{2}} = 0$

3. $\lim\limits_{x \to 4} (x-4)^{\frac{1}{2}}$

In order for a limit to exist, both its left and right hand limits must exist AND be the same. In this case, the left hand limit does not exist, therefore the limit cannot exist either.

$\lim\limits_{x \to 4} (x-4)^{\frac{1}{2}} = \text{DNE}$

4. $\lim\limits_{x \to 4^-} (x-4)^{\frac{1}{3}}$

Even though x is approaching 4 from the left, resulting in negative numbers under the radical, it does not matter because the radical is odd, not even, and you can take odd radicals of negative numbers.

$\lim\limits_{x \to 4^-} (x-4)^{\frac{1}{3}} = (4-4)^{\frac{1}{3}} = 0$

5. $\lim\limits_{x \to 4^+} (x-4)^{\frac{1}{3}}$

Since the radical is odd, it doesn't matter what number is under the radical, the radical will exist. Hence, we just plug in the specified input and solve as above.

$\lim\limits_{x \to 4^+} (x-4)^{\frac{1}{3}} = (4-4)^{\frac{1}{3}} = 0$

6. $\lim\limits_{x \to 4} (x-4)^{\frac{1}{3}}$

In order for a limit to exist, both its left and right hand limits must exist and be the same. In this case, we solved for both the left and right hand limits of this limit above. They both existed, and they were both the same. Hence, the limit is equal to them.

$\lim\limits_{x \to 4} (x-4)^{\frac{1}{3}} = 0$

7. $\lim\limits_{x \to 0^+} \dfrac{|x|}{x}$

In solving this problem, we note that upon initially plugging in the specified number, we get the absolute value of zero divided by zero. We can't factor this, so we must graph.

$\lim\limits_{x \to 0^+} \dfrac{|x|}{x} = \dfrac{|0|}{0}$

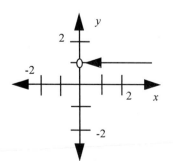

Plugging in values for x greater than 0, since x is approaching 0 from the right, we obtain the adjacent graph. The result is clear: The right hand limit is equal to 1.

8. $\lim\limits_{x \to 0^-} \dfrac{|x|}{x}$

As in problem Example 7 above, we get the absolute value of zero divided by zero, which is not factorable. Hence, we must graph again. When we do this, plugging in values less than 0 since we are approaching 0 from the left, we see that the left hand limit equals -1.

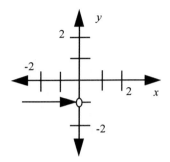

$$\lim\limits_{x \to 0^-} \dfrac{|x|}{x} = \dfrac{|0|}{0}$$

9. $\lim\limits_{x \to 0} \dfrac{|x|}{x}$

This limit does not exist because its left and right hand limits are not the same--even though both the right hand limits do exist.

$$\lim\limits_{x \to 0} \dfrac{|x|}{x} = \text{DNE}$$

10. $\lim\limits_{x \to 2^+} \dfrac{(x^2 - 2)(x + 1)}{(x + 5)}$

As we noted in the introduction to this section, the normal rules of limits apply to one-sided limits. Examples 10 and 11 illustrate this. Basically, it's just "plug 'n chug".

$$\lim\limits_{x \to 2^+} \dfrac{(x^2 - 2)(x + 1)}{(x + 5)} = \dfrac{\left[(2)^2 - 2\right](2 + 1)}{(2 + 5)} = \dfrac{6}{7}$$

11. $\lim\limits_{x \to 10^-} 15$

$$\lim\limits_{x \to 10^-} 15 = 15$$

Additional Problems   2.2
Find the limits if they exist.

1. $\lim\limits_{x \to 20^+} (x - 20)^{\frac{1}{2}}$

3. $\lim\limits_{x \to 20} (x - 20)^{\frac{1}{2}}$

5. $\lim\limits_{x \to 2^-} \dfrac{(3x^2 - 2)}{(x - 1)}$

2. $\lim\limits_{x \to 20^-} (x - 20)^{\frac{1}{2}}$

4. $\lim\limits_{x \to 1^+} (6x^3 + x + 7)$

6. $\lim\limits_{x \to 137^+} 2$

7. $\lim\limits_{x\to 1^-} \dfrac{(x-1)^2}{x+1}$

13. $\lim\limits_{x\to 7^-} \dfrac{7-x}{|7-x|}$

19. $\lim\limits_{x\to 27} x^{\frac{1}{3}}$

8. $\lim\limits_{x\to 1^+} \dfrac{(x+1)^2}{x-1}$

14. $\lim\limits_{x\to 7} \dfrac{7-x}{|7-x|}$

20. $\lim\limits_{x\to 3^-} \dfrac{1}{(6-2x)}$

9. $\lim\limits_{x\to 10^-} \dfrac{|x-10|}{x-10}$

15. $\lim\limits_{x\to 2^+} \dfrac{(x^2-4)}{x-2}$

21. $\lim\limits_{x\to 0^+} \left[(6-2x)^{\frac{1}{2}} + x^{\frac{1}{2}}\right]$

10. $\lim\limits_{x\to 10^+} \dfrac{|x-10|}{x-10}$

16. $\lim\limits_{x\to 0^+} \dfrac{\left(\sqrt{5+x}-1\right)}{x}$

22. $\lim\limits_{x\to 9^+} \dfrac{\left[(x-9)^2\right]^{\frac{1}{2}}}{x-9}$

11. $\lim\limits_{x\to 10} \dfrac{|x-10|}{x-10}$

17. $\lim\limits_{x\to 27^-} x^{\frac{1}{3}}$

23. $\lim\limits_{x\to 9^-} \dfrac{\left[(x-9)^2\right]^{\frac{1}{2}}}{x-9}$

12. $\lim\limits_{x\to 7^+} \dfrac{7-x}{|7-x|}$

18. $\lim\limits_{x\to 27^+} x^{\frac{1}{3}}$

24. $\lim\limits_{x\to 9} \dfrac{\left[(x-9)^2\right]^{\frac{1}{2}}}{x-9}$

2.3 Trigonometric Function Limits

Trigonometric function limits are a special topic in the study of limits. They have their own set of theorems, and solving them often involves much trigonometric manipulation. If you can remember your basic trigonometry, you should have an easy time with this topic. Here's how we go about solving them:

To solve Trig limit problems, plug the specified number into the function.

1. If you get a number, that is the limit of the function.

2. If you get a constant divided by zero, the function has no limit for the specified number.

3. If you get zero divided by zero, factor the top and bottom, cancel, and plug in again repeating the process. If you can't factor, and in these problems you usually won't be able too, use one of the trig theorems below to simplify and solve.

Trig Limit Theorems

1. $\displaystyle\lim_{x \to 0} \sin x = 0$ 2. $\displaystyle\lim_{x \to 0} \cos x = 1$

3. $\displaystyle\lim_{x \to 0} \frac{\sin x}{x} = 1$ 4. $\displaystyle\lim_{x \to 0} \frac{1 - \cos x}{x} = 0$

Examples   2.3

1. $\displaystyle\lim_{x \to 0} \frac{\sin^2 x}{x}$

$$\lim_{x \to 0} \frac{\sin^2 x}{x} = \lim_{x \to 0} \left[\frac{\sin x}{x} \cdot \sin x \right] = (1)(0) = 0$$

The trick here, as with most of these trig limit problems, is to rewrite the function in a way that "exposes" one of the theorems. Here, we factor a sin x out of the function. In so doing, we see the first theorem above and can readily solve by plugging in the specified number. Remember: the sine of zero is zero.

2. $\displaystyle\lim_{x \to 0} \frac{\left(7x^3 + \sin x\right)}{x}$

$$\lim_{x \to 0} \frac{\left(7x^3 + \sin x\right)}{x} = \lim_{x \to 0} \left[\frac{7x^3}{x} + \frac{\sin x}{x} \right]$$

$$= \lim_{x \to 0} \left[7x^2 + \frac{\sin x}{x} \right] = 7(0)^2 + 1 =$$

In this problem we simply break the original function into two pieces and simplify before plugging in the specified number to solve. Remember, the goal is to expose one of the theorems above.

3. $\displaystyle\lim_{x \to a} \frac{\sin 19x}{2x}$

$$\lim_{x \to a} \frac{\sin 19x}{2x} = \lim_{x \to a} \left[\left(\frac{19}{19} \right) \left(\frac{\sin 19x}{2x} \right) \right]$$

$$= \lim_{x \to a} \left[\frac{19 \sin 19x}{(19)(2x)} \right]$$

$$= \lim_{x \to a} \left[\left(\frac{19}{2} \right) \frac{\sin 19x}{19x} \right] = \frac{19}{2}(1) = \frac{19}{2}$$

There are 2 tricks in this problem:

(1) Realizing that the third theorem above is general, and

(2) Figuring out what to factor to take advantage of this fact.

In this case, we need a 19x in the denominator to account for the sin19x. We obtain this by multiplying the top and bottom of the fraction by 19/19, then factoring out the 19 on top and the 2 on the bottom.

4. $\lim\limits_{x \to 0} \tan x$

$$\lim\limits_{x \to 0} \tan x = \lim\limits_{x \to 0} \frac{\sin x}{\cos x} = \frac{\sin 0}{\cos 0} = \frac{0}{1} = 0$$

In this case we don't need to take advantage of the theorems because merely rewriting tangent in terms of sine and cosine and then plugging in the specified number enables us to solve directly. REMEMBER: Always look for the easy way first!

5. $\lim\limits_{x \to 0} \dfrac{\sin 5x}{x}$

$$\lim\limits_{x \to 0} \frac{\sin 5x}{x} = \lim\limits_{x \to 0} 5\left(\frac{\sin 5x}{5x}\right) = 5(1) = 5$$

This example is just like Example #3 above, in that we must multiply the function by a fraction equal to 1 to solve- -in this case 5/5.

6. $\lim\limits_{x \to 0} \dfrac{\tan^2 x}{2x \sin x}$

$$\lim\limits_{x \to 0} \frac{\tan^2 x}{2x \sin x} = \lim\limits_{x \to 0} \frac{\left(\dfrac{\sin^2 x}{\cos^2 x}\right)}{2x \sin x} = \lim\limits_{x \to 0} \frac{1}{2}\left(\frac{\sin x}{x}\right)\left(\frac{1}{\cos^2 x}\right)$$

$$= \frac{1}{2}(1)(1) = \frac{1}{2}$$

As a general rule, whenever you see a tangent, convert it to its sine and cosine form. From there, it's a lot easier to figure out how to solve the problem. In this case, we can readily find an above theorem, then just plug in and solve once this conversion is done.

7. $\lim\limits_{x \to 1} \dfrac{\sin(x^7 - 1)}{x^7 - 1}$

$$\lim\limits_{x \to 1} \frac{\sin(x^7 - 1)}{x^7 - 1} = \frac{\sin 0}{0} \quad = 1$$

This is a tricky problem. To solve it, we note that after plugging in the specified number, zero, we obtain the same result as occurs in theorem number 3 above. The similarity holds, and the result is the same. The limit is equal to 1.

8. $\lim\limits_{x \to 0} \dfrac{\left(\sqrt{5 - \cos 3x} - 2\right)}{x}$

Step #1: Conjugate: $\left(\sqrt{5 - \cos 3x} + 2\right)$

In this problem the only way to solve is by using the conjugate technique. The goal of the conjugate technique here is to enable us to "expose" one of the above theorems. See Example #16, from Examples 2.1 to review the conjugate technique if necessary.

Step #2: $\displaystyle\lim_{x\to 0}\left[\frac{\left(\sqrt{5-\cos 3x}-2\right)}{x}\right]\times\left[\frac{\sqrt{5-\cos 3x}+2}{\sqrt{5-\cos 3x}+2}\right]$

$\displaystyle =\lim_{x\to 0}\frac{5-\cos 3x-4}{x\left(\sqrt{5-\cos 3x}+2\right)}=\lim_{x\to 0}\left(\frac{1-\cos 3x}{3x}\right)\left(\frac{3}{\sqrt{5-\cos 3x}+2}\right)$

Step #3: $\displaystyle\lim_{x\to 0}\left(\frac{1-\cos 3x}{3x}\right)\left(\frac{3}{\sqrt{5-\cos 3x}+2}\right)=(0)\left(\frac{3}{\sqrt{5-\cos 3x}+2}\right)=0$

Additional Problems   2.3

Find the limits if they exist.

1. $\displaystyle\lim_{x\to 0}\frac{\tan x}{x}$

2. $\displaystyle\lim_{x\to 0}\frac{\sin^2 3x}{x\sin 3x}$

3. $\displaystyle\lim_{x\to 0}\frac{x}{\sin x}$

4. $\displaystyle\lim_{x\to 0}\frac{x+\sin x}{\sin x}$

5. $\displaystyle\lim_{x\to 0}\frac{1-\cos 9x}{3x}$

6. $\displaystyle\lim_{x\to 0}\frac{\sin 3x}{18x}$

7. $\displaystyle\lim_{x\to 0}\frac{x(1-\cos x)}{x+5}$

8. $\displaystyle\lim_{x\to 0}\frac{x\cot x}{\cos x}$

9. $\displaystyle\lim_{x\to 0}\frac{6\cos x-6}{17x}$

10. $\displaystyle\lim_{x\to 0}x\csc x$

11. $\displaystyle\lim_{x\to 0}\left[x(1+\tan^2 3x)(\sin 3x)\right]$

12. $\displaystyle\lim_{x\to 0}\frac{\sin 2x}{x\cos x}$

13. $\displaystyle\lim_{x\to 0}\frac{1}{2}\left(\frac{\sin(x+\pi)+\sin(x-\pi)}{x\cos\pi}\right)$

14. $\displaystyle\lim_{x\to 0}\frac{\sin^2 x}{x^2}$

15. $\displaystyle\lim_{x\to 0}\frac{(2-\cos x)^{\frac{1}{2}}-1}{x}$

16. $\displaystyle\lim_{x\to 0}\frac{\sin 17x}{\sin 67x}$

17. $\displaystyle\lim_{x\to 0} \frac{\sin 310x}{438x}$

18. $\displaystyle\lim_{x\to 0} \frac{1-\cos 37x}{1093x}$

19. $\displaystyle\lim_{x\to 0} \frac{\cos 19x}{\cos 2x}$

20. $\displaystyle\lim_{x\to 0} \frac{\sin(3-x)}{3-x}$

21. $\displaystyle\lim_{x\to 0} \frac{1-\cos 5x}{19x}$

22. $\displaystyle\lim_{x\to 0} \frac{4x^3 + x^2 \sin 3x}{x^3}$

23. $\displaystyle\lim_{x\to 0} \frac{\cos\left(x + \dfrac{3\pi}{2}\right)}{x}$

2.4 Continuity

Now that we've covered the basic essentials of limits, it's time to get on with some applications. With this in mind, we end this chapter with a discussion of *continuity*, and begin the next with a discussion of *derivatives*--both applications of limits. Specifically, in this section we're going to be talking about the *continuity of functions*. What do we mean, continuity of functions? Well, here we are concerned with knowing whether a function is continuous at a certain point, and also whether it is continuous on a closed interval, e.g., [a, b].

Just as a quick reminder of what we're talking about when we say continuous, the function $f(x)$, shown in Fig. 2.6, is continuous throughout the real numbers and therefore continuous on any interval of real numbers. The function $g(x)$, on the other hand, shown in Fig. 2.7, is not continuous everywhere throughout the real numbers. It is discontinuous at $x = 1$ and therefore said to be discontinuous on any interval containing $x = 1$.

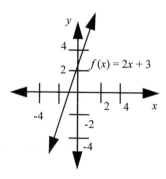

Fig. 2.6

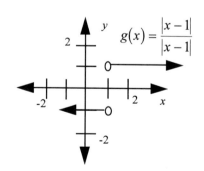

Fig. 2.7

Now in these examples you probably could figure all of this out just by observation, however, many cases will not be this obvious. Therefore, we employ the following theorems:

Theorems for Determining the Continuity of a Function

1. A function f is continuous at a number a if
       i. $f(a)$ is defined
       ii. $\lim\limits_{x \to a} f(x)$ exists
       iii. $\lim\limits_{x \to a} f(x) = f(a)$

2. A function f is continuous on a closed interval $[a,b]$ if:
       i. f is continuous throughout (a,b)
       ii. $\lim\limits_{x \to a^{+}} f(x)$ exists
       iii. $\lim\limits_{x \to a^{-}} f(x) = f(b)$

3. If the functions f and g are continuous at a point a, then $(f - g)$, $(f + g)$, $(f \cdot g)$ are continuous at a. If $g \neq 0$, then $(f \div g)$ is also continuous at a.

4. Polynomial functions are continuous throughout the real numbers.

Except for a few conditions, there is nothing new to learn here. You already know the techniques needed to determine if these conditions exist, so solving these upcoming continuity problems should be nothing more than advanced "plug and chug". Relax and enjoy as you go through the examples on the following pages.

Examples   **2.4**

1.   Is $f(x) = \begin{cases} 2 \text{ if } x = 3 \\ -1 \text{ if } x \neq 3 \end{cases}$ continuous at 3?

Clearly, from observation we can see that the function is discontinuous at x = 3. We can prove this observation, though, using the theorems outlined in this section.

$[1]$ $f(3) = 2$ $\therefore$ $f(3)$ is defined

$[2]$ $\lim\limits_{x \to 3} f(x) = -1$ $\therefore$ the limit exists

$[3]$ $\lim\limits_{x \to 3} f(x) \neq 3$ $\therefore$ the function is discontinuous at $x = 3$

2.   Is $f(x) = -2x + \sqrt{5x - 6}$, continuous at $x = 2$?

In this case you can't easily determine the answer just by looking at the function, hence the rules we've learned come in handy.

$[1]$ $f(2) = -4 + \sqrt{10 - 6} = -4 + 2 = -2$

$[2]$ $\lim\limits_{x \to 2} \left(-2x + \sqrt{5x - 6}\right) = -4 + 2 = -2$

$[3]$ $\lim\limits_{x \to 2} f(x) = f(2)$ $\therefore$ $f(x)$ is continuous at $x = 2$.

3.   Is $f(x) = \dfrac{x + 1}{x - 1}$ continuous at $x = 1$?

In this example, since the first condition for continuity is not met--the function is not defined at the point in question--the function is discontinuous at the point.

$[1]$ $f(1) = \dfrac{1 + 1}{1 - 1} = \dfrac{2}{0}$ $\therefore$ DNE $\therefore$ $f(x)$ is discontinuous at $x = 1$.

4.   Is $f(x) = \dfrac{x^2 - 1}{x + 1}$ continuous at $x = -1$?

In this example we stumble immediately on the first step. Upon plugging in our desired number into the function, we get zero divided by zero, which tells us nothing. Hence, we must factor, simplify, then try again.

$[1]$ $f(-1) = \dfrac{1 - 1}{-1 + 1} = \dfrac{0}{0}$ $\therefore$ Factor

Factoring: $\dfrac{x^2-1}{x+1} = \dfrac{(x-1)(x+1)}{x+1} = x-1$

$f(-1) = (x-1) = -1-1 = -2$

[2] $\lim\limits_{x\to-1} (x-1) = -2$

[3] $\lim\limits_{x\to-1} f(x) = f(-1)$ $\therefore$ $f(x)$ is continuous at $x = -1$

5. Is $f(x) = x^2 + 2$ continuous on $[1,\ 9]$?

[1]   From observation, we see there are no points in
this interval that can cause $f(x)$ to be discontinuous.

[2] $\lim\limits_{x\to 1^+} (x^2 + 2) = 1 + 2 = 3$

$f(1) = 1 + 2 = 3$

$\therefore$ $\lim\limits_{x\to 1^+} f(x) = f(1)$

[3] $\lim\limits_{x\to 9^-} (x^2 + 2) = 81 + 2 = 83$

$f(9) = 81 + 2 = 83$

$\therefore$ $\lim\limits_{x\to 9^-} f(x) = f(9)$

6. Is $f(x) = \begin{cases} x^2 + 1, \text{ if } x \neq -3 \\ -5, \text{ if } x = -3 \end{cases}$ continuous on $[-4,\ 4]$?

[1]   We see that the function is discontinuous at $x = -3$
and, therefore, discontinuous on $(-4,\ 4)$. Therefore,
$f(x)$ is discontinuous on $[-4,\ 4]$.

7. Is $f(x) = \begin{cases} x^2 + 1, & \text{if } x \neq -3 \\ -5, & \text{if } x = -3 \end{cases}$ continuous on $[-3, 1]$?

   [1]   We see no discontinuity on $(-3, 1)$

   [2] $\lim\limits_{x \to -3^+} (x^2 + 1) = (-3)^2 + 1 = 10$

         $f(-3) = -5$

   $\therefore \lim\limits_{x \to -3^+} f(x) \neq f(-3) \; \therefore \; f(x)$ is discontinuous on this interval.

Additional Problems   2.4
In the following problems, determine if the function is continuous at the given point a.

1. $f(x) = 2$ $a = 17$

2. $f(x) = 2x + 1$ $a = 0$

3. $f(x) = \dfrac{x^2}{x-1}$ $a = 1$

4. $f(x) = \dfrac{x^2 - 9}{x}$ $a = 1$

5. $f(x) = \dfrac{x^2 - 4}{x + 2}$ $a = -2$

6. $f(x) = \dfrac{\sqrt{x-2}}{2x-1}$ $a = 1$

7. $f(x) = \sqrt{x - 3}$ $a = 3$

8. $f(x) = \dfrac{x^2 - 25}{x + 5}$ $a = -5$

9. $f(x) = (x - 7)^{\frac{1}{3}}$ $a = 34$

10. $f(x) = (x - 7)^{\frac{1}{2}}$ $a = 7$

11. $f(x) = \cos x$ $a = 1$

12. $f(x) = \tan x$ $a = 1$

13. $f(x) = \dfrac{\cos x}{\sin x + 3}$ $a = 0$

14. $f(x) = \dfrac{(x - 1)^2}{x - 2}$ $a = 2$

In the following problems determine if the function is continuous on the given interval.

15. $f(x) = \dfrac{1}{x}$ $[-1, 1]$

16. $f(x) = 7x^2 + 3$ $[3, 1093]$

17. $f(x) = \dfrac{|x-1|}{x-1}$ $[-2, 3]$

18. $f(x) = \dfrac{|x-1|}{x-1}$ $[2, 8]$

19. $f(x) = \dfrac{\sqrt{x-3}}{2x^3 - 9}$ $[2, 30]$

20. $f(x) = \dfrac{(x+8)^2}{x+2}$ $[-1, 1]$

In the following problems, find the domain of the function.

21. $f(x) = \dfrac{1}{x}$

22. $f(x) = \dfrac{x^2 - 3x + 1}{x - 1}$

23. $f(x) = \dfrac{(x-1)^2}{x^2 - 13x + 2}$

24. $f(x) = \cos x$

25. $f(x) = 2x - 3$

26. $f(x) = \dfrac{x}{(x-3)^{\frac{1}{3}}}$

27. $f(x) = \dfrac{|x+3|}{x+3}$

Derivatives

As you will recall, in the first chapter of this book we reviewed functions because calculus is the branch of mathematics devoted to the study of three mathematical operations--limits, derivatives, and integrals--which are used on functions to determine a variety of things. In the second chapter we learned about the first of these operations, limits, and noted that their main importance was in helping to define the other two. Not wanting to remain in the realm of the theoretical too long, we quickly move on in this chapter to the second operation, the derivative, which is used the world over in such professions as business, science, engineering, and medicine. Enjoy!

Surprised at the apparent widespread usefulness of the derivative? Wondering what it is about the derivative that makes it appealing to such a diverse set of professions? Well, to answer this question, answer another: In what professions do you think it's important to be able to determine rates? What sorts of people would be interested in determining how fast their retirement fund was growing, how fast the economy was growing or shrinking, how fast the new drug killed bacteria, how quickly a 737 passenger jet can change directions without shearing off its wings, what speed a satellite needs to attain in order to remain in orbit at a certain altitude, how quickly the knee moves with respect to the hip after a hip replacement, how quickly an air bag inflates upon impact, or how quickly a particular medicine can diffuse through a cell membrane? Get the idea? Spend some time coming up with some more examples on your own. With a little thought, you'll generate quite a list, and at the same time answer your question about why derivatives are so appealing.

3.1 The Basis of the Derivative

As you may have already guessed from the introduction to this chapter, *a derivative is simply a mathematical operation that determines rate, specifically the rate of change of a function's points* (by points we mean the (x, y) ordered-pairs which are frequently represented graphically on a Cartesian coordinate system--see Chapter 1). The emphasis here is on the phrase *rate of change*. Derivatives are rate determiners--that is all. How they make their rate determinations, and their basis, is the focus of this section.

Think back for a moment to your years of algebra. What concept did you learn about that had to do with rates? That's right, slope. Specifically, the slope of a straight line. You remember the classic "rise over run" formula: $(y_2 - y_1)/(x_2 - x_1)$. Using this formula, you were able to calculate the steepness, or slope, of a straight line--in other words, how fast the line is rising or falling. It's not real difficult, though you're probably wondering how it relates to derivatives since derivatives are concerned with determining the rate of change of a function's points--points that don't always fall in a straight line. Well, if we modify the formula a bit *using limits* [DING! DING!], we find that the slope concept stays intact and serves our need quite nicely.

Before we go any further with this discussion, we need to clarify our objective. From what has just been said, you might have gotten the impression that in asking to find the rates of change of a function's points, we are asking to find the slopes of those points. This, however, would be impossible to find since points, by definition, have no size, shape or direction. *What we are really looking for is a measure of how one point of a function changes relative to the other points around it.* One way to obtain such a measure is to determine the slopes of the lines tangent to these points (See Fig. 3.1).

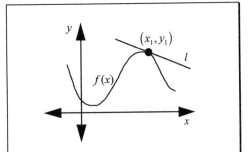

Fig. 3.1: A graph of the function, $f(x)$, showing the line tangent to the point (x_1, y_1). The slope of this tangent line tells us how (x_1, y_1) is changing relative to the points around it.

So, finding the slopes of the lines tangent to the points of a function is what a derivative does. Essentially, the derivative operation takes a function and modifies it, changing it to a new function. This new function has the same domain as the original function, but its output (its range) is different. The outputs of the new function represent the slopes of the lines tangent to the original function (See Fig. 3.2).

This is a little tricky, so let's step back for a minute and put everything into perspective. We have a function. That function, when given an input, say x,

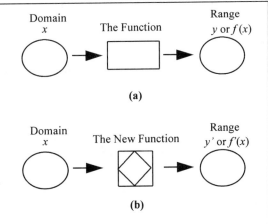

Fig. 3.2: Picture **a** shows the basic mechanics of a function. Picture **b** shows the what happens after a derivative operates on the function. Note: The domain of the new function is the same, but the range--its output--is different.

outputs a value y. Grouping these two numbers as an ordered-pair, we have a point that can be plotted on a Cartesian coordinate system. If we plot enough of the function's points on the graph, we can connect them so that the function is represented by some sort of line--just like the function $f(x)$ is in Fig. 3.1. Now, we ask ourselves, what is the rate of change of the function at the point (x_1, y_1)? To find out, we take the derivative of the function, getting a new function. We input x_1 into this new function and get an output y_1'. y_1' tells us the rate of change at the point (x_1, y_1)--i.e., the slope of the line tangent to (x_1, y_1). Similarly, if we wanted to find the rate of change of another point in the function now, say (x_3, y_3), all we have to do is plug x_3 into the new function to get our answer.

Getting back to the idea of slopes now, we would like to know how a derivative does its job of finding the slopes of the tangent lines to the points of a function. We will explain this by example, so look over Fig. 3.3 carefully before we start.

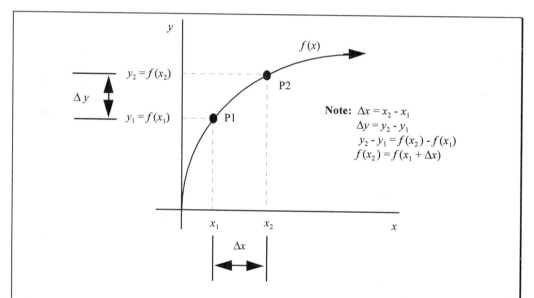

Fig. 3.3: This is a graph of some function, $f(x)$, depicted on your basic Cartesian coordinate system. There are two points of interest, P1 and P2, with coordinates (x_1, y_1), and (x_2, y_2) respectively. Note that by using function notation we can write the same things several different ways.

Suppose we want to use a derivative to find the rate of change of the point P1 of the function $f(x)$ in Fig. 3.3 above. Here's what the derivative does:

First, the derivative picks another point, P2, that is close to P1. Then it calculates the difference between x_2 and x_1, and y_2 and y_1. The differences are called *delta x* (Δx), and *delta y* (Δy) respectively, where the term *delta* means *the exact change in*. Next, the derivative moves P2 infinitely close to P1 by causing the x value of P2 to *approach* that of P1 [DING DING!!]. Once P2 is infinitely close to P1, the derivative calculates the corresponding Δx and Δy values (which are very small at this stage) and plugs them into the classic slope formula to get the answer--the value of the slope of the line tangent to P1. Mathematically, this process is represented as follows:

$$\Delta x = x_2 - x_1$$

$$\Delta y = y_2 - y_1 = f(x_2) - f(x_1) = f(x_1 + \Delta x) - f(x_1)$$

$$\text{slope} = \frac{\text{rise}}{\text{run}} = m$$

You can determine delta x and delta y just by looking at Fig. 3.3 and taking advantage of all the equivalencies.

$$\therefore \quad \text{Derivative} = m = \lim_{\Delta x \to 0} \frac{\Delta y}{\Delta x}$$

$$= \lim_{\Delta x \to 0} \frac{f(x_1 + \Delta x) - f(x_1)}{x_2 - x_1}$$

Once delta y and delta x are known, coming up with an expression for the derivative is just a matter of plugging them into the slope formula and tacking on a limit as specified in our above discussion.

Note that the limit is used to move P2 infinitely close to P1, and that it's just attached to the classic slope formula of rise over run. Essentially, we have modified this old slope formula so that it can work on a microscopic level.

So, from the above derivation we see the two basic mathematical forms in which the derivative is expressed. At this point we will introduce the formal notation of the derivative.

Notation: Rather than use m to represent derivatives, we use one of these three forms:

$$D_x[f(x)], \quad \frac{d[f(x)]}{dx}, \quad f'(x)$$

--where $f(x)$ is the original function. Note that $f'(x)$ is pronounced, f prime of x. The other two forms have no special pronunciation.

That's all there is to the derivative. It's really just a glorified application of slopes with limits. When you get a problem, as you will see in the following examples, you simply do the following:

Method for solving derivatives:

1. Plug the given function into the derivative formula:

$$f'(x) = \lim_{\Delta x \to 0} \frac{f(x + \Delta x) - f(x)}{\Delta x}$$

2. Solve as you would a normal limit.

As you can probably imagine, using the method outlined above to solve for the derivative of a function has the potential to lead to some very heinous calculations because we're forced to deal with limits. Fear not, the rest of the chapter is not like this. There is an easier way to calculate derivatives which we will learn about in the next section. First, though, we have to make sure we understand the hard way as it shows us the basis of the derivative.

Examples   3.1

1. $f(x) = 2x$

$$f'(x) = \lim_{\Delta x \to 0} \frac{f(x + \Delta x) - f(x)}{\Delta x}$$

$$= \lim_{\Delta x \to 0} \frac{2(x + \Delta x) - 2x}{\Delta x}$$

$$= \lim_{\Delta x \to 0} \frac{2x + 2\Delta x - 2x}{\Delta x}$$

$$= \lim_{\Delta x \to 0} \frac{2\Delta x}{\Delta x} = 2$$

As specified in the methods box on page 41, we begin this problem by plugging the given function into the derivative formula. The only difficult thing about this is the is the f(x+Δx) expression. What this expression is telling you to do is to take the function and put an (x+Δx) wherever you see an x. For the f(x) part, you simply put your function in.

In this particular example, our function is 2x, so it's simple. We put in the 2(x+Δx) for the f(x+Δx) part, and 2x for the f(x) part. We then multiply and simplify.

2. $f(x) = 3x - 5$

$$f'(x) = \lim_{\Delta x \to 0} \frac{[3(x + \Delta x) - 5] - (3x - 5)}{\Delta x}$$

$$= \lim_{\Delta x \to 0} \frac{3x + 3\Delta x - 5 - 3x + 5}{\Delta x}$$

$$= \lim_{\Delta x \to 0} \frac{3\Delta x}{\Delta x} = 3$$

In this example things get a little more complicated because there's more to the function, but the basic idea remains--you pull out the derivative formula and plug away. Just remember, for the f(x+Δx) part, wherever you see an x in the original function, you put an (x+Δx).

3. $f(x) = x^2 + 5x + 10$

No surprises here. The only thing of note is that the function is a tad more complex, so you really have to be careful when you plug in (x+Δx).

$$f'(x) = \lim_{\Delta x \to 0} \frac{\left[(x + \Delta x)^2 + 5(x + \Delta x) + 10\right] - \left[x^2 + 5x + 10\right]}{\Delta x}$$

$$= \lim_{\Delta x \to 0} \frac{\left[x^2 + 2x\Delta x + \Delta x^2 + 5x + 5\Delta x + 10\right] - x^2 - 5x - 10}{\Delta x}$$

$$= \lim_{\Delta x \to 0} \frac{2x\Delta x + \Delta x^2 + 5\Delta x}{\Delta x}$$

$$= \lim_{\Delta x \to 0} \frac{\Delta x(2x + \Delta x + 5)}{\Delta x} = \lim_{\Delta x \to 0} (2x + \Delta x + 5) = 2x + 0 + 5 = 2x + 5$$

4. $f(x) = 17$

An easy problem that sometimes throws people at this stage of the game. What we have here is a function with no variables. Hence, the f(x+Δx) is the same as the f(x), 17.

$$f'(x) = \lim_{\Delta x \to 0} \frac{17 - 17}{\Delta x}$$

$$= \lim_{\Delta x \to 0} \frac{0}{\Delta x} = \lim_{\Delta x \to 0} 0 = 0$$

5. $f(x) = (x + 2)^2$

$$f'(x) = \lim_{\Delta x \to 0} \frac{\left[(x + \Delta x) + 2\right]^2 - (x + 2)^2}{\Delta x}$$

$$= \lim_{\Delta x \to 0} \frac{\left[x^2 + 2x\Delta x + 4x + 4\Delta x + \Delta x^2 + 4\right] - x^2 - 4x - 4}{\Delta x}$$

$$= \lim_{\Delta x \to 0} \frac{2x\Delta x + 4\Delta x + \Delta x^2}{\Delta x}$$

$$= \lim_{\Delta x \to 0} \frac{\Delta x(2x + 4 + \Delta x)}{\Delta x} = 2x + 4 + 0 = 2x + 4$$

6. $f(x) = x^{-2}$

After going through the last five examples, when you see this one it looks like a walk in the park. You make your initial substitution as always, and then figure it's over. That is, until you start to do the algebra...

$$f'(x) = \lim_{\Delta x \to 0} \frac{(x + \Delta x)^{-2} - x^{-2}}{\Delta x}$$

$$= \lim_{\Delta x \to 0} \frac{\left[\dfrac{1}{(x + \Delta x)^2} - \dfrac{1}{x^2} \right]}{\Delta x}$$

$$= \lim_{\Delta x \to 0} \frac{\left[\dfrac{x^2 - (x + \Delta x)^2}{x^2 (x + \Delta x)^2} \right]}{\Delta x}$$

$$= \lim_{\Delta x \to 0} \frac{\left[\dfrac{x^2 - \left(x^2 + 2x\Delta x + \Delta x^2 \right)}{x^2 (x + \Delta x)^2} \right]}{\Delta x}$$

$$= \lim_{\Delta x \to 0} \frac{\left[\dfrac{-\Delta x (2x + \Delta x)}{x^2 (x + \Delta x)^2} \right]}{\Delta x}$$

$$= \lim_{\Delta x \to 0} \frac{-2x - \Delta x}{x^2 (x + \Delta x)^2} = \frac{-2x - 0}{x^2 (x + 0)^2} = -\frac{2x}{x^4} = -2x^{-3}$$

As you can see, things start getting nasty in a hurry. Before it's all said and done, you end up with one whale of an algebra mess on your hands.

This is a good example for you to work through for 2 reasons: (1) Your teacher/professor will probably pull something like this on a homework/quiz/exam, and (2) it's an excellent algebra review.

If you can crank through this problem, your life in calculus is going to be far easier than you ever imagined.

Additional Problems   3.1

Given the function $f(x)$, find $f'(x)$.

1. $f(x) = 3x$

2. $f(x) = -7x$

3. $f(x) = 19x + 96$

4. $f(x) = -3x - 4$

5. $f(x) = 1$

6. $f(x) = -2472$

7. $f(x) = x^2 + x$

8. $f(x) = 3x^2 - 7x + 2$

9. $f(x) = (2x - 7)^2$

10. $f(x) = 2x^{-2}$

3.2 The Easy Way

After going through the rather extreme mental gymnastics in the last section, you're probably wondering just how necessary derivatives are. Fortunately, a while back some young scholars who felt the same way came up with a set of rules that greatly simplifies the whole process. Memorize these rules. Circle them in crayon if you have to. Whatever it takes, memorize them. Here they are:

Easy Rules for Finding Derivatives:

1. $D_x(c) = 0$, where c is a constant. Example: $f(x) = 3$, then $f'(x) = 0$.

2. $D_x(x^n) = nx^{n-1}$, where n is any real number but zero. Example: $f(x) = x^3$, $f'(x) = 3x^2$

3. $D_x[cf(x)] = cD_x[f(x)]$, where c is a constant. Example: $f(x) = 3x^3$, $f'(x) = 9x^2$

4. $D_x[f(x) + g(x)] = D_x[f(x)] + D_x[g(x)]$. Example: $h(x) = 3x^2 + 4x^8$, $h'(x) = 6x + 32x^7$

5. $D_x[f(x) - g(x)] = D_x[f(x)] - D_x[g(x)]$. Example: $h(x) = 3x^2 + 4x^8$, $h'(x) = 6x - 32x^7$

6. $D_x\left[\dfrac{f(x)}{g(x)}\right] = \dfrac{g(x)D_x[f(x)] - f(x)D_x[g(x)]}{[g(x)]^2}$, where $g(x) \neq 0$

   Example: $h(x) = \dfrac{5x + 5}{2x}$, then $h'(x) = \dfrac{(2x)(5) - (5x + 5)(2)}{(2x)^2}$

(Note: A simple way to remember this division rule is the following saying: "Low d high, minus high d low, over which the denominator squared must go").

7. $D_x(\sin x) = \cos x$

8. $D_x(\cos x) = -\sin x$

Examples   3.2

1. $f(x) = 2$

In this problem we have a function that is a constant. According to the rules above, the derivative of a constant is always zero.

$f'(x) = 0$

2. $f(x) = x^7$

Problems #2 and #3 are examples of functions with just variables raised to some power. To find the derivative of these types of functions, a copy of the exponent hops down in front, then on the top you subtract one from it.

$f'(x) = 7x^{7-1} = 7x^6$

For example, in Problem #2 the 7 hops down in front, then we subtract 1 from it on top to get a new exponent value, 6. This is the derivative.

In Problem #3, the 5 hops down in front, then we subtract 1 from it on top to get a new exponent value, 4.

3. $f(x) = x^5$

$f'(x) = 5x^{5-1} = 5x^4$

4. $f(x) = 19x^2$

In this example we have a function with a constant multiplied by a variable raised to a power. To find the derivative here, the exponent hops down, gets multiplied by the constant, then 1 is subtracted from the exponent on top.

$f'(x) = (19)(2)x^1$

$= 38x$

So, here we have the 2 hopping down, getting multiplied by the 19, followed by 1 being subtracted from 2 on top.

5. $f(x) = 5x$

This example is the same as #4 above, except our variable is raised to the first power. Hence, 1 hops down, gets multiplied by 5, then 1 is subtracted from 1 on top. This results in a new exponent of 0, and as you will recall from algebra, anything raised to the 0 power is just 1. Hence, our answer is just 5.

$f'(x) = (5)(1)x^0$

$= 5$

6. $f(x) = 2x^3 + 5x + 1$

This example differs from the previous ones in that we have several functions within the function that need to be worked on. To solve, all we do is find the derivative of each individually, then put them together.

$f'(x) = 2(3)x^2 + 5x^0 + 0$

$= 6x^2 + 5$

7. $f(x) = 12x^3 - 2x^4 + x - 19$

Example #7 is just like Example #6, except here we have some minus signs. No big deal, just find the derivatives taking into account this minor distraction.

$f'(x) = 12(3)x^2 - 2(4)x^3 + 1x^0 - 0$

$= 36x^2 - 8x^3 + 1$

8. $f(x) = \dfrac{7x}{x^2 + 1}$

Here's an example of the quotient rule from above. Basically, we have two functions being divided. To solve, remember the handy saying:

$f'(x) = \dfrac{(x^2 + 1)(7) - (7x)(2x)}{(x^2 + 1)^2}$

$= \dfrac{7x^2 + 7 - 14x^2}{(x^2 + 1)^2}$

$= \dfrac{7 - 7x^2}{(x^2 + 1)^2}$

"Low d high, minus high d low, over which the denominator squared must go" (d stands for derivative--example, low d high means you take the lower function and multiply it by the derivative of the top one).

9. $f(x) = 7x^{-3}$

Remember how difficult it was to find the derivative of a variable raised to a negative power using the modified slope formula in the last section? Well, using the handy rules from above, look how easy it is--the exponent hops down, then 1 is subtracted from the exponent. Simple!

$f'(x) = (7)(-3)x^{-4}$

$= -21x^{-4}$

10. $f(x) = \cos x + 3\sin x - x^2 + 5$

One final example with sines and cosines. There's no difference in these types of problems, you just have to remember that the derivative of cosine is negative sine, and that the derivative of sine is cosine.

$f'(x) = -\sin x + 3\cos x - 2x$

Additional Problems   3.2

Given the function $f(x)$, find $f'(x)$.

1. $f(x) = 1996$

2. $f(x) = -21$

3. $f(x) = -\dfrac{1}{9}$

4. $f(x) = \dfrac{\pi}{2}$

5. $f(x) = x^3$

6. $f(x) = x^{597}$

7. $f(x) = 3x^2$

8. $f(x) = 7x^5$

9. $f(x) = -8x^2$

10. $f(x) = 3x^2 + 7x^5$

11. $f(x) = 11 + 2x + 7x^3$

12. $f(x) = -11x^5 + x^3 + 3x^2$

13. $f(x) = x - 1$

14. $f(x) = x^2 - x - 1$

15. $f(x) = 13x^2 - 7x^6 + \dfrac{3942}{13} + x$

16. $f(x) = \dfrac{1}{x+1}$

17. $f(x) = \dfrac{x^3 + x}{7x + 1}$

18. $f(x) = \dfrac{8x^2 + x + 1}{x^3}$

19. $f(x) = x^{-1}$

20. $f(x) = x^{-6}$

21. $f(x) = x^{-19}$

22. $f(x) = 2x^{-3}$

23. $f(x) = -3x^{-7}$

24. $f(x) = 20x^{-2}$

25. $f(x) = x^{\frac{1}{2}}$

26. $f(x) = x^{\frac{1}{3}}$

27. $f(x) = x^{-\frac{1}{7}}$

28. $f(x) = x^{-\frac{1}{2}}$

29. $f(x) = 3x^{-\frac{1}{3}}$

30. $f(x) = \sin x$

31. $f(x) = \cos x$

32. $f(x) = 10\cos x + 13\sin x + x^3 - x^{-2}$

3.3 Increments and Differentials

Suppose we have a function and we want to know what the exact change in its output will be when we change its input by some exact amount. For example, say we want to input the number 2 and the number 2.01--the goal here is to find out exactly how much the output differs. Clearly, one way to determine this would be to simply plug both numbers into the function, record their output, then subtract the results. This, however, can be cumbersome at times, and if we're asked to determine some generic formula that will work for any set of inputs, this method is useless. There are better ways.

You'll recall from our earlier discussions that we use the delta (Δ) notation when describing the exact change in a variable. In our present situation, if x is the independent variable and y the dependent variable, then what our above question translates into is, "Given Δx, what is Δy?" One way to calculate this that works every time is to use the following formula which we say back in section 3.1:

$$\Delta y = f(x + \Delta x) - f(x)$$

Unfortunately, while this formula works, the calculations that result from it can be miserable. Hence, we develop the following shorthand method which is good for *approximating* changes in y based on exact changes in x:

First, remember our modified slope formula:

$$f'(x) = \lim_{\Delta x \to 0} \frac{\Delta y}{\Delta x}$$

What we want to do here is estimate Δy given Δx. Well, if Δx is small--close to zero-- then we can drop the limit from this equation without introducing too much error:

$$f'(x) \approx \frac{\Delta y}{\Delta x}$$

Rearranging terms, we get:

$$\Delta x \cdot f'(x) \approx \Delta y$$

Now, notice the approximation sign. In order to get rid of it, we introduce *differentials*, which are defined as follows:

$dx = \Delta x$ *Note: dy and dx are called differentials.*

$dy = \Delta x\, f'(x)$, thus $f'(x) = dy/dx$

Note that *dy* does not equal Δy, it's only approximately equal to it. However, *dy* is a very good approximation of Δy, and much easier to calculate. Why? Because we don't have to use that formula for Δy that is so cumbersome. We can just take a quick derivative and we're done. In the following examples you'll see just how good *dy* is.

Examples   **3.3**

In the following problems, find Δy, *dy*, and Δy-*dy*.

1. $y = f(x) = 2x^2 + 7$

In this example we're given a function and told that the independent variable x, changes some exact amount, Δx. (Actually, we're not told this, but it is implied). We want to determine how this affects the dependent variable y--i.e., how much y changes, Δy.

$$\Delta y = f(x + \Delta x) - f(x)$$
$$= \left[2(x + \Delta x)^2 + 7\right] - \left[2x^2 + 7\right]$$
$$= 2x^2 + 4x\Delta x + 2\Delta x^2 + 7 - 2x^2 - 7$$
$$= 4x\Delta x + 2\Delta x^2$$

In order to determine this, we first calculate Δy the hard way. Then we approximate Δy by calculating dy. Notice that it is much easier to calculate dy.

$$f'(x) = \frac{dy}{dx} = 4x$$
$$\therefore\ dy = (4x)dx$$

To see how much error is introduced when we do the approximation, we subtract dy from Δy. Notice that the error is very small-- equal to only $2\Delta x$ squared. If Δx is small, this error is negligible. Also, don't forget, dx=Δx

$$\therefore\ \Delta y \text{-} dy = 4x\Delta x + 2\Delta x^2 - (4x)dx$$
$$= 2\Delta x^2$$

In the following problems, given $\Delta x = 0.02$ and $x = 10$, find $\Delta y - dy$.

2. $y = f(x) = 3x^2 + 5$

This problem is exactly the same as the last one, except this time we're given numbers to plug in to illustrate how much error we'd get in a real world situation.

$$\Delta y = \left[3(x + \Delta x)^2 + 5\right] - \left[3x^2 + 5\right]$$

$$= 3x^2 + 6x\Delta x + 3\Delta x^2 + 5 - 3x^2 - 5$$

$$= 6x\Delta x + 3\Delta x^2 = 6(10)(0.02) + 3(0.02)^2 = 1.2012$$

To solve this problem, we begin by finding Δy, then take the derivative, dy/dx, and rearrange to get dy. We then subtract dy from Δy to get our error. As you can see, the error is not very big-- even when Δx is not that small.

$$\frac{dy}{dx} = f'(x)$$

$$= 6x$$

$$\therefore \ dy = 6x \ dx = 6(10)(0.02) = 1.2$$

$$\Delta y - dy = 1.2012 - 1.2 = 0.0012$$

3. Suppose you're going to buy some new carpet for your room. If your room's floor is in the shape of a perfect square, and you estimate the sides to be 20 feet long give or take an tenth of a foot, how much carpet should you buy?

Let x = length of a wall

Let A = area of floor

$$A = x^2 = (20)^2 = 400 \text{ ft}^2$$

Your error in measurement of the lenght of the wall is, however, 0.1 ft. So,

Let $\Delta x = dx = 0.1$

Thus, the exact length of the wall is between, $x + dx$ and $x - dx$

To quickly approximate how much error in area calculation this error in our measurement has brought on, we use differentials:

$$\frac{dA}{dx} = 2x$$

$$dA = 2x \ dx = 2(20)(0.1) = 4.0 \text{ ft}^2$$

No big deal. Even if your wall measurement is off, it doesn't throw the area calculation off too much. If you buy around 405 square feet you should be fine. Suppose, however, that this carpet is very expensive and you want to know exactly how much the error in your wall measurement affects your area calculation. In that case, this quick estimate won't do. We have to use increments:

$$\Delta A = (x + \Delta x)^2 - x^2$$
$$= x^2 + 2x\Delta x + \Delta x^2 - x^2$$
$$= 2x\Delta x + \Delta x^2 = 2(20)(0.1) + (0.1)^2 = 4.01 \text{ ft}^2$$

As you can see, the quick approximation was very good, only being off 0.01 ft^2.

Additional Problems   **3.3**

In the following problems, find Δy, dy, and Δy - dy.

1. $f(x) = 4x + 3$

3. $f(x) = 9x^2 - 3x + 19$

2. $f(x) = 4x^2 + 3$

4. $f(x) = x^2 + 5x - 1$

In the following problems, given $x = 5$ and $\Delta x = 0.01$, find Δy - dy.

5. $f(x) = 10x^2 + 2$

6. $f(x) = 3x^2 + 2x + 2$

Solve the following story problems.

7. Suppose your neighbors down the hall saw what a great job you did carpeting your room, so they ask you to carpet a section of their's. If the section of interest is 10 foot by 10 foot, with a possible error of 0.01 feet in this estimate on each side, approximate how much this error in measurement affects the area calculation, then determine exactly how much it does.

8. After finishing with your little side job, you wonder on out to the field across from the dorms to see what's going on. Unfortunately, as soon as you get there someone sets off a monstrous smoke bomb. The smoke blast shoots straight up into the air covering a spherical volume of 5 million feet cubed, with an error in that estimate of 100,000 feet cubed. Running at your top speed you can be 108 feet away from ground zero when the smoke hits ground at the maximum radius corresponding to the maximum volume. Use differentials to determine if you make it, then increments.

3.4 The Chain Rule

Now that we're done with that little aside, we can get on with the business of learning how to calculate derivatives. If you look back to section 3.2, you'll notice that none of the rules presented there are much good for calculating the more complex functions constantly encountered in calculus. Since nobody wants to go back to the using limits as in section 3.1, we must press on and learn some new techniques that will handle these more complex functions for us. The rules which we are about to learn--there are three of them--are collectively referred to as the chain rule. Since these rules are best taught by example, that is how we will proceed.

Rule #1: To take the derivative of two *interrelated* functions, for example, $f(x) = \sin 5x$, where we say the *sine* and the *5x* are interrelated, you:

(1) Take the derivative of each function.
(2) Multiply those derivatives together.

In this particular instance we take the derivative of *sin 5x* and get *cos 5x*. We then take the derivative of *5x* and get *5*. Multiplying them together, we get $f'(x) = 5 \cos 5x$, which is the correct answer. (Note: At this stage the only interrelated functions you're going to see are those of the trigonometric persuasion. Hence, when you see a sine or a cosine beware).

Rule #2: In order to take the derivative of a function raised to some power, say *n*, you:

(1) Take the function and raise it to the *n-1* power.
(2) Multiply that whole quantity by *n*.
(3) Multiply that whole quantity by the derivative of the function.

Mathematically, these three steps are written as: $D_x[f(x)]^n = n[f(x)]^{n-1} D_x[f(x)]$

Translation: The derivative of the function f(x) raised to the n power, equals n times the function f(x) raised to the n-1 power, times the derivative of the function.

For example, if we want to take the derivative of the function, $f(x) = (\sin 5x)^2$, using the above guidelines we get: $f'(x) = 2 (\sin 5x) (5 \cos 5x)$.

Rule #3: To take the derivative in cases where you have more than one function multiplied together, for example $f(x) = (7x + 3)^2(11x - 2)^3$, you:

(1) Take the derivative of the first term and multiply it by the second term.
(2) Take the derivative of the second term and multiply it by the first term.
(3) Add the two quantities together.

Carrying on with the example outlined in Rule #3, the answer would be:

$$f'(x) = 14(7x+3)(11x-2)^3 + 33(11x-2)^2(7x+3)^2$$

The step-by-step procedure for obtaining this result is as follows:

1. Derivative of the first term, $(7x+3)^2$, is: $2(7x+3)(7) = 14(7x+3)$

   Multiplying the derivative of the first term by the second term:

   $$14(7x+3)(11x-2)^3$$

2. Derivative of the second term, $(11x-2)^3$, is: $3(11x-2)^2(11) = 33(11x-2)^2$

   Multiplying the derivative of the second term by the first term:

   $$33(11x-2)^2(7x+3)^2$$

3. Adding the two quantities from #1 and #2, we get the answer:

   $$f'(x) = 14(7x+3)(11x-2)^3 + 33(11x-2)^2(7x+3)^2$$

Examples   **3.4**

Given $f(x)$, find $f'(x)$.

1. $f(x) = \sin 10x$

   Our first chain rule example is with interrelated functions, in this case sine and 10x. To solve, we take the derivative of each, then multiply them together. Recall that the derivative of sinx is cosx, and the derivative of 10x is 10.

   $$f'(x) = 10\cos 10x$$

2. $f(x) = \cos 7x$

   As in the previous example, we have interrelated functions. This time cosine and 7x. Remember, the derivative of cosx is -sinx, and the derivative of 7x is 7.

   $$f'(x) = -7\sin 7x$$

3. $f(x) = (x+1)^3$

   In this example we have a function raised to a power. Hence, we follow the guidelines in Rule #2: We raise the function to the 3-1 power, we multiply it by 3, and then take the derivative of it, which equals 1, and multiply the whole quantity by that.

   $$f'(x) = 3(x+1)^2(1) = 3(x+1)^2$$

4. $f(x) = (7x^2 - 2x + 1)^4$

   Very similar to the previous example, the only difference between the two problems is that the function in this case is a bit more complex. Still, the basic procedure remains as is outlined in Rule #2.

   $$f'(x) = 4(7x^2 - 2x + 1)^3(14x - 2)$$

4. $f(x) = (2x^2 - 1)^{-2}$

We tossed this example in just to make sure you really understood the second rule--remember, we're subtracting 1 from the exponent. Hence, if we have a negative exponent to begin with, it gets more negative. In this case, from -2 to -3.

$f'(x) = -2(2x^2 - 1)^{-3}(4x)$

5. $f(x) = 3(\cos 2x)^4$

This is a tricky problem. In order to solve it you need to use both Rule #1 and Rule #2. You start out using Rule #2 because we have a function raised to a power. However, in completing the final step in Rule #2, calculating the derivative of the function, we find it necessary to employ Rule #1.

$f'(x) = 3(4)(\cos 2x)^3(-2\sin 2x)$

$= -12(\cos 2x)^3(-2\sin 2x)$

6. $f(x) = x^2(2x + 5)^3$

In this example we must employ Rule #3. Fairly simple, we just chug through the steps as outlined in the text on pages 52 and 53.

$f'(x) = 2x(2x + 5)^3 + x^2(3)(2x + 5)^2(2)$

7. $f(x) = (x^3 + 1)(x^3 - 1)^5$

This problem is the same type as the previous one, just a little more involved since the functions are more complex.

$f'(x) = 3x^2(x^3 - 1)^5 + (x^3 + 1)\left[5(x^3 + 1)^4(3x^2)\right]$

Additional Problems   3.4

Given $f(x)$, find $f'(x)$.

1. $f(x) = \sin 3x$

2. $f(x) = \cos 11x$

3. $f(x) = \sin 8x$

4. $f(x) = -\sin 29x$

5. $f(x) = -3\cos 30x$

6. $f(x) = 7\cos 7x$

7. $f(x) = (2x^2 + 7x)^3$

8. $f(x) = (x + 1)^{39}$

9. $f(x) = 5(x^3 - 9x^2 + 3x + 11)^5$

10. $f(x) = (3x^4 - 2x)^{-5}$

11. $f(x) = (3x^2 + 5x)^{-\frac{1}{2}}$

18. $f(x) = x^4(x+1)^3$

12. $f(x) = (\sin 10x)^3$

19. $f(x) = 15(x^2 + 2x)^3(2x+1)$

13. $f(x) = (\cos 3x)^5$

20. $f(x) = (x^2 + 2)^2(\cos 5x)^3$

14. $f(x) = 59(\cos 2x)^{-1}$

21. $f(x) = \left(\dfrac{x^2}{x+1}\right)(2x^3 + 2x + 1)^3$

15. $f(x) = -6(\sin 10x)^{11}$

22. $f(x) = 5x^2(x+1)^3(x^2 + 2x^3 + 5x)^5$

16. $f(x) = (3x^3 + 2x + 1)(x+1)^3$

23. $f(x) = \left(\dfrac{\cos 10x}{\sin^2 3x}\right)^5$

17. $f(x) = (x^4 - 1)^5(2x^2 - x)^7$

24. $f(x) = (2x+1)^5(x^2 + 5x)^4(-3x^4 - 1)^2$

3.5 Implicit Functions: Finding Their Derivatives

What is an implicit function? Sounds like one of those things you were suppose to have learned about in Algebra II, but have tried desperately to forget. Well, fear not. Implicit functions are nothing more than functions that are not written in the standard $y = f(x)$ form. For example, in the equation, $3x - 4 + y = 2$, y is said to be an implicit function of x because you could easily rewrite the equation in the standard $y = f(x)$ form. In this case, rewriting the equation, we get, $y = f(x) = -3x + 6$. In other words, an implicit function is one that implies that y is a function of x without expressly stating it.

In most cases that we'll be running into, the implicit functions will not be this obvious--in other words, it won't be easy to isolate y by itself on one side of the equation. This is not important, though, because in most instances the problems we'll be facing will begin by telling us that y is an implicit function. Our job is to find the derivative of y: y'. Therefore, all you need to know is what an implicit function is, and of course, how to solve for its derivative. You've got the former down if you've gotten this far. Here's how you accomplish the latter:

Assume that $x^4 - y^2 + 2y = 7$ is an implicit function of y, such that $y = f(x)$ and that you want to solve for the derivative y'. To do this you:

1. Take the derivative of each with respect to x - -(y' is the derivative of y with respect to x).

   $$4x^3 - 2yy' + 2y' = 0$$

2. Solve for y':

   $$y'(-2y + 2) = -4x^3$$

   $$y' = \frac{-4x^3}{(-2y + 2)}$$

3. Plug $f(x)$ in for y for the aesthetic quality.

   $$y' = \frac{-4x^3}{(-2f(x) + 2)}$$

Some of you may be feeling a little queasy at this point by the derivative taking of the y with respect to x--in other words, y'. If this is the case, then note that all we're doing here is taking advantage of the chain rule. It just so happens that instead of calling the derivative of y $f'(x)$, we're calling it y'. Nothing wrong with that.

Examples   3.5
For the following problems, y is an implicit function. Find y'.

1. $2x^3 + y^3 - y = 8$

   To solve this problem, we just do what we did in the above example: We take the derivative of each term, with respect to x.

   $$6x^2 + 3y^2 y' - y' = 0$$

   $$y'(3y^2 - 1) = -6x^2$$

   Once the derivatives are taken, we then rearrange terms to get y' by itself on one side of the equation.

   $$y' = \frac{-6x^2}{(3y^2 - 1)}$$

2. $x^2 y^3 + y^5 = 2x + 1$

   This problem is a little more complex because the derivatives are a little more difficult to get. In order to solve for them, we need to utilize the chain rules.

   $$(2xy^3 + 3x^2 y^2 y') + 5y^4 y' = 2$$

   $$y'(3x^2 y^2 + 5y^4) = 2 - 2xy^3$$

   Again, as in the last problem, once the derivatives are found, it's just a matter of rearranging terms.

   $$y' = \frac{2 - 2xy^3}{3x^2 y^2 + 5y^4}$$

3. $7x^3 - y + xy = 23$

A fairly simple example. We included this because it shows slightly different terms which, though not difficult to take the derivative of, are somewhat confusing if you've never seen them before.

$$21x^2 - y' + (y + xy') = 0$$

$$y'(x-1) = -21x^2 - y$$

$$y' = \frac{-21x^2 - y}{x - 1}$$

Additional Problems   3.5
For the following problems, y is an implicit function. Find y'.

1. $y + 2xy + 7x^2 = 394$

2. $x^2 + y^2 = 36$

3. $19x^2 + y^7 + y^4 x^3 = 3x^9 + 2x^2 + 1$

4. $x + y + x^2 - y^2 - x^3 y^4 = 0$

5. $x^3 = \dfrac{y - x}{x + y}$

6. $y^3 = 2x \cos y$

7. $(y + 5)^8 = xy^2 + y^4$

3.6 Higher Order Derivatives
This will undoubtedly be the shortest section in this book as the topic is simple. A higher order derivative is just the taking of derivatives of a function as many times as the order indicates. For example, a second order derivative represented by $f''(x)$, indicates that you take the derivative of $f(x)$ twice. For example, say $f(x) = 2x^2$, then $f'(x) = 4x$, and $f''(x) = 4$. If you wanted to take the third order derivative of this function, $f'''(x)$, you would get $f'''(x) = 0$. As you can see, each subsequent derivative is taken from the equation resulting from the derivative before. More examples follow.

Examples   3.6

1. $f(x) = 7x^3 + 2x + 1, \ f'(x) = ?, \ f''(x) = ?, \ f'''(x) = ?$

$$f'(x) = 21x^2 + 2$$
$$f''(x) = 42x$$
$$f'''(x) = 42$$

2. $f(x) = x^{-2} + 2x$, $f'(x) = ?$, $f''(x) = ?$, $f'''(x) = ?$

$f'(x) = -2x^{-3} + 2$

$f''(x) = 6x^{-4}$

$f'''(x) = -24x^{-5}$

3. $f(x) = 5\cos 5x$, $f'(x) = ?$, $f''(x) = ?$, $f'''(x) = ?$

$f'(x) = -25\sin 5x$

$f''(x) = -125\cos 5x$

$f'''(x) = 625\sin 5x$

4. $f(x) = (2x^2 + 4x + 1)^3$, $f'(x) = ?$, $f''(x) = ?$

$f'(x) = 3(2x^2 + 4x + 1)^2 (4x + 4)$

$\qquad = (12x + 12)(2x^2 + 4x + 1)^2$

$f''(x) = (12)(2x^2 + 4x + 1)^2 + (12x + 12)\left[2(2x^2 + 4x + 1)(4x + 4)\right]$

Additional Problems   3.5

1. $f(x) = x^8$, $f''(x) = ?$, $f'''(x) = ?$

2. $f(x) = 2x^4$, $f''(x) = ?$, $f'''(x) = ?$

3. $f(x) = \sin x$, $f''(x) = ?$, $f'''(x) = ?$

4. $f(x) = 8x^4 - x^{-2} + x - 309$
$\qquad f''(x) = ?$, $f'''(x) = ?$

5. $f(x) = (x + x^2 + 5)^{\frac{1}{3}}$, $f''(x) = ?$

6. $f(x) = (x + 1)^2 (2x^3 - 2)^4$, $f''(x) = ?$

7. $f(x) = 2\cos 3x$, $f''(x) = ?$, $f'''(x) = ?$

8. $f(x) = -5(\sin 2x)^2$, $f''(x) = ?$

Derivative Applications: Max-Mins and Rates

In the last chapter we learned about the second of the three mathematical operations that comprise calculus--the derivative. In this chapter we go a step further, expanding our discussion to applications of derivatives. You should enjoy this! These applications are, for the most part, very practical.

By now you have a pretty good understanding of derivatives. You know that derivatives are mathematical operations used to determine the rate of change of a function's points, you know that they are made up of limits and the old slope formula, and you know how to use them on many different types of functions. As of yet, however, you do not know how to apply them to the real world, which is essential if they're to be of any real value to you. Hence, in this chapter we begin to show you some practical applications of derivatives.

4.1 Max-Mins: The Basic Idea

The first application of derivatives that we're going to show you is that of the max-mins. What are max-mins? Well, when we say max-mins we're referring to the maximum and minimum output values of a function--specifically, the points at which these maximum and minimum output values occur. In this section we're going to learn how to use derivatives to find these points. "Why would we want to do that?" you ask. Because there are a lot of instances in the real world where being able to quickly determine the maximum and minimum points of a function is quite useful. For example:

Real World Application
In business, companies are always trying to maximize their profits. This is not always easy, but with the help of careful record keeping and derivatives, it's possible. How? Well, if a company can express its cost of manufacturing a particular product as a function of production, derivatives can be used on that function to determine how many goods need to be produced to maximize profit. Examples in section 4.4 illustrate this.

Bet you didn't think this calculus would come in handy for running a business! And that's just one example. Here's another:

Real World Application
Engineers are problem solvers. That's their main function. Often, in industry, this skill is focused on producing goods in as cost effective manner as possible. For example, an engineer is faced with the problem of designing a battery case for a new battery. Unfortunately, the material required for this particular battery is incredibly expensive. Hence, the engineer is told to design the case so that it will be able to contain a certain volume while at the same time using as little material as possible. How can the engineer solve this problem? That's right, you guessed it--using max-mins. Examples in section 4.4 illustrate how this is done.

We're not done yet. Here's an example for you pre-med types:

Real World Application
Doctors often do studies on patients and drugs to determine the effectiveness of the drugs. For example, suppose a young patient is suffering from A.D.D. (Attention Deficit Disorder). In an effort to combat the disorder, a doctor prescribes the drug *Elox* and keeps track of the various doses given to the patient. The doctor also keeps track of the young patients grades in school, and after a period of time, figures out what two functions represent the doses and the grades. In order to determine if there is a correlation between grades and drug dosages, an important first piece of information would be the maximum and minimum points of both of these functions. Hence, derivatives would be of value-- much easier to use then pouring through the reams of collected data.

You get the idea. There's a lot of use out there for max-mins. So, let's learn more about them.

The basic idea behind max-mins is that *any function that is continuous on some closed interval [a, b] has a maximum and a minimum point occurring in that interval at least once.* These maximum and minimum points can occur at the endpoints of the interval and/or at points within the interval. The greatest maximum point on an interval is called the *absolute max*, and the smallest minimum point is called the *absolute min*. Other maximum and minimum points within the interval are referred to as *local max's* and *local min's*. See Fig. 4.1*a*.

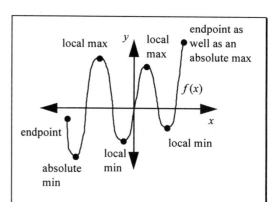

Fig. 4.1 a: A graph of the function $f(x)$, over a given interval.

Notice how the designation of the maximum and minimum points within the interval as being either absolute or local is entirely dependent on the chosen interval. In other words, if you change the interval, you may change the designation. See Fig. 4.1 *b*.

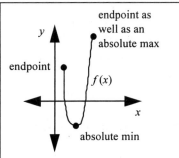

So how exactly do we go about finding these maximum and minimum points of a function? We use the following two-step process:

1. We begin by identifying the critical points-- i.e., all the points of a function that may possibly be maximum or minimums.

Fig 4.1 b: Same function, $f(x)$, but this time with a smaller interval. Note how the designation of "absolute" and "local" maximum and minimum point depends on the interval.

2. Run either the First Derivative Test or the Second Derivative Test to determine which of the critical points that lie within the interval of interest are maximums or minimums.

In this section we'll learn how to find the critical points. In the next, we'll learn how to do the First and Second Derivative tests--as well as how to use the information it gives us to graph functions.

Finding the Critical Points

At first glance, it may seem a rather difficult task to find all of a function's critical points given only the formula for the function. However, if you look back to Figs 4.1 *a* and *b*, you'll notice that all the critical points (except those at the endpoints) occur in places where the slopes of the lines tangent to the given function are zero--i.e., horizontal. See Fig. 4.2.

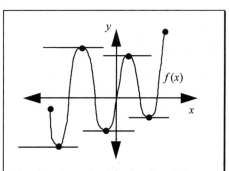

Fig. 4.2: A graph of the function $f(x)$, over a given interval. Notice that all the critical points within the interval occur in places where the slopes of the lines tangent to the function are zero--i.e., horizontal.

DING!! DING!! This last statement should be ringing a bell for you. Why? Because in the last chapter we learned about derivatives. What do derivatives tell us? They tell us what the slopes of the lines tangent to the points of a function are. Hence, all we have to do in order to find the critical points of a function is find the points where the derivative of the function equals zero. This is not hard. The following three-step method demonstrates how it's done.

Rules for finding critical points:

1. If an interval is specified, plug the x-values into the function to determine their corresponding y values, then write out the ordered pairs--these are both critical points. For example, if you're given the function $f(x) = 2x$, and you're given the interval [-1, 5], we get:

$f(-1) = 2(-1) = -2$
$f(5) = 2(5) = 10$, thus: (-1, -2) and (5, 10) are two of the critical points.

2. Take the first derivative of the function, set it equal to zero, and solve for all possible values of x. These values of x are referred to as *critical numbers*. Plug them into the function to obtain their corresponding y values--the resulting ordered pairs are critical points.

3. There is one more place where critical points can occur--at values of x where $f'(x)$ does not exist, but $f(x)$ does. For example,

$$f(x) = x^{\frac{1}{2}}; f(0) = (0)^{\frac{1}{2}} = 0$$

$$f'(x) = \frac{1}{2}x^{-\frac{1}{2}} = \frac{1}{2}\left(\frac{1}{x^{\frac{1}{2}}}\right); f'(0) = \frac{1}{2}\left(\frac{1}{(0)^{\frac{1}{2}}}\right) = \frac{1}{2}\left(\frac{1}{0}\right) = \text{Does not exist.}$$

Here, a critical point occurs at $x = 0$. To get the coordinate of the point simply plug x into the function and get its corresponding y value.

The following examples should make the process of finding critical points crystal clear.

Examples   **4.1**
Find the critical points.

1. $f(x) = 2x^2 + 1$

To find the critical points of this function-i.e., the possible maximum and minimum points, we follow the step-by-step process outlined above.

[1] No interval given, thus no endpoints so we skip step #1.

Since no interval is specified there are no endpoints and thus nothing to do as far as step #1 is concerned. Hence, we skip it and move on to step #2.

[2] $f'(x) = 4x$

$\qquad 4x = 0$

$\qquad x = 0$

$\qquad f(0) = 2(0)^2 + 1 = 1$ $\therefore$ critical point: $(0,1)$

In step #2, we take the first derivative of the function, set the result equal to zero, and solve. The resulting number, in this case 0, is the x-coordinate of a critical point. To get its corresponding y-coordinate, we simply plug it into the function.

[3] Since $f'(x)$ exists for all real numbers, no critical points here.

In step #3 we find no critical points as f'(x) exists everywhere.

$\therefore$ The critical points for this function are: $(0,1)$

Thus there's only 1 critical point for this function.

2. $f(x) = \dfrac{2}{3}x^3 - \dfrac{9}{2}x^2 - 35x + 12$; for $[-2,10]$

In this example an interval is specified, hence step #1 must be completed--not a difficult step, just plugging and chugging.

[1] $f(-2) = \dfrac{2}{3}(-2)^3 - \dfrac{9}{2}(-2)^2 - 35(-2) + 12 = 69.3$

$\quad f(10) = \dfrac{2}{3}(10)^3 - \dfrac{9}{2}(10)^2 - 35(10) + 12 = -121.3$

$\therefore$ critical points at: $(-2,69.3)$ and $(10,-121.3)$

[2] $f'(x) = 2x^2 - 9x - 35$

$\qquad 2x^2 - 9x - 35 = 0$

$\qquad (2x + 5)(x - 7) = 0$

$\qquad x = -\dfrac{5}{2} = -2.5$, and $x = 7$

Note: -2.5 is out of the interval, hence we ignore it.

In step #2, we again take the derivative, set it equal to zero, and solve. Note that one of the critical points is outside of the specified interval, thus we ignore it--we don't care what's going on outside the interval.

$f(7) = \dfrac{2}{3}(7)^3 - \dfrac{9}{2}(7)^2 - 35(7) + 12 = -224.8$

$\therefore$ critical point at: $(7,-224.8)$

[3] Since $f'(x)$ exists for all real numbers, no critical points here.

This time we have three critical points of interest--remember, we don't list the one that falls outside the interval.

$\therefore$ The critical points for this function in the interval, $[-2,10]$ are: $(-2,69.3)$, $(7,-224.8)$ and $(10,-121.3)$

3. $f(x) = x^{\frac{1}{3}}$

[1] No interval given, $\therefore$ skip step.

[2] $f'(x) = \dfrac{1}{3}x^{-\frac{2}{3}}$

$\dfrac{1}{3}x^{-\frac{2}{3}} = 0$

$\dfrac{1}{3}\left(\dfrac{1}{\sqrt[3]{x^2}}\right) = 0$ $\Leftarrow$ No value of x satisfies this

$\therefore$ No critical point here.

[3] $f'(x)$ does not exist at $x = 0$, however $f(x)$ does:

$f(0) = (0)^{\frac{1}{3}}$

$f'(0) = \dfrac{1}{3}(0)^{-\frac{2}{3}} = \dfrac{1}{3}\left(\dfrac{1}{(0)^{\frac{2}{3}}}\right) = \dfrac{1}{0} =$ Does not exist

$\therefore$ critical point at: $(0,0)$

$\therefore$ The critical points for this function are: $(0,0)$

Additional Problems

In the following problems, find the critical points.

1. $f(x) = 3x^2 + 2$

2. $f(x) = -7x^2 + 3$

3. $f(x) = \dfrac{2}{3}x^3 + 4x^2 + 8x + 1996$

4. $f(x) = 3x^2 + \dfrac{23}{2}x^2 - 12x + 1$

5. $f(x) = 2x + 7$

6. $f(x) = 7 - 5x$

7. $f(x) = 10x^2 + 5x - 13$

8. $f(x) = 3x^2 - 10x + 2$

9. $f(x) = \dfrac{x+1}{x}$

10. $f(x) = \dfrac{5x^2 + 2x + \dfrac{1}{2}}{2x + 1}$

11. $f(x) = x^{\frac{1}{2}}$

12. $f(x) = 3x^{\frac{2}{3}} + x + 1$

4.2 Determining Max-Min Points and Graphing Functions

OK, we know how to solve for a function's critical points--the possible maximum and minimum points of the function--but how exactly do we determine if these points are in fact maximums or minimums? Well, as we alluded to in the last section, we can use the First Derivative test, or another test known as the Second Derivative Test--both of which we're going to learn about in this section. Additionally, besides being able to tell us whether or not a function's critical points are maximums, minimums, or neither, these tests can be used to help us graph functions.

Before we get into the specifics of these tests, however, we need to quickly explain the idea of increasing and decreasing intervals:

A function is said to be *increasing* if, as you move along the *x*-axis from left to right, its *y*-values (its output), increase, and subsequently said to be *decreasing* if, as you move along the *x*-axis from left to right, its *y*-values (its output), decrease. Fig. 4.3 illustrates this.

As you're about to see, this increasing/decreasing concept is one of the keys to both the First and Second Derivative Tests. We now move on to the First.

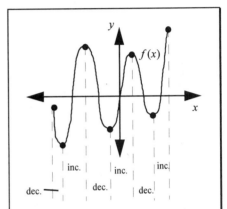

Fig. 4.3: A graph of the function $f(x)$, showing where the function is said to be *increasing* and *decreasing*.

The First Derivative Test

The purpose of the First Derivative Test is to determine which, if any, of a function's critical points are maximums or minimums. Here's how it works:

1. Find the critical points using the method described in Section 4.1.

2. Locate any *x*-values that cause the function $f(x)$ to be discontinuous or not exist.

3. Divide the *x*-axis into intervals, where the boundaries of the intervals are the endpoints, critical numbers, and discontinuities of the function. See Fig. 4.3.

4. For each of these intervals, determine whether the function is increasing, decreasing, or continuous. You do this by plugging a number from each interval into the function, and then looking at the resulting sign: (+) = increasing, (-) = decreasing.

5. For each critical number, determine if its corresponding critical point is a maximum, a minimum, or neither. The way you do this is by going back to your *x*-axis with the marked intervals. If you see:

I. An *increasing* interval followed by a *decreasing* interval, the critical point between them is a maximum.

II. A *decreasing* interval followed by an *increasing* interval, the critical point between them is a minimum.

III. *Two increasing* intervals in a row, or *two decreasing* intervals in a row, then the point between them is neither a maximum nor a minimum point.

Examples   4.2.1
Find the locations of the maximum and minimum points of the given functions using the First Derivative Test. *Note: Graphs of these functions are located in corresponding Examples 4.2.2.*

1. $f(x) = 3x^3 + 4x^2 - x + 1, \ [-3,3]$

$[1]$ Find the critical points:

(1) $f(-3) = 3(-3)^3 + 4(-3)^2 - (-3) + 1 = -4$

$f(3) = 3(3)^3 + 4(3)^2 - (3) + 1 = 115$

∴ critical points at: $(-3,-41), \ (3, \ 115)$

The first part of this First Derivative Test is just like the last section--we're finding the critical points.

(2) $f'(x) = 9x^2 + 8x - 1$

$9x^2 + 8x - 1 = 0$

$(9x - 1)(x + 1) = 0$

$x = \dfrac{1}{9}, \ x = -1$

$f\left(\dfrac{1}{9}\right) = 3\left(\dfrac{1}{9}\right)^3 + 4\left(\dfrac{1}{9}\right)^2 - \left(\dfrac{1}{9}\right) + 1 = 0.94$

$f(-1) = 3(-1)^3 + 4(-1)^2 - (-1) + 1 = 3$

As you can see, there are four critical points for this function for the given interval.

∴ critical points at: $(-1,3), \ \left(\dfrac{1}{9}, 0.94\right)$

(3) $f'(x)$ exists throughout the real numbers, thus no critical points here.

∴ The critical points for this function for the interval $[-3,3]$ are:

$(-3,-41), \ (-1,3), \ \left(\dfrac{1}{9}, 0.94\right), (3, \ 115)$

[2] $f(x)$ is continuous everywhere

[3] $[-3,-1], \left[-1,\dfrac{1}{9}\right], \left[\dfrac{1}{9}, 3\right]$

[4] $f'(-2) = 9(-2)^2 + 8(-2) - 1 = 19 \;\therefore\; (+) \;\therefore\;$ increasing

$\quad f'(0) = 9(0)^2 + 8(0) - 1 = -1 \;\therefore\; (-) \;\therefore\;$ decreasing

$\quad f'(1) = 9(1)^2 + 8(1) - 1 = 16 \;\therefore\; (+) \;\therefore\;$ increasing

[5] $x = -1,$ increasing $\rightarrow$ decreasing $\therefore$ max

$\quad x = \dfrac{1}{9},$ decreasing $\rightarrow$ increasing $\therefore$ min

$\therefore$ Looking at the critical points then, we see that:

$\quad (-3,-41)$ is an absolute min

$\quad (-1,3)$ is a local max

$\quad \left(\dfrac{1}{9}, 0.94\right)$ is a local min

$\quad (3,115)$ is an absolute max

2. $f(x) = x^4 - 3x^3 + 10$

[1] Find the critical points.

(1) No interval given, no critical points there.

(2) $f'(x) = 4x^3 - 9x^2$

$\quad 4x^3 - 9x^2 = 0$

$\quad x^2(4x - 9) = 0$

$\quad x = 0, \; x = \dfrac{9}{4}$

$\quad f(0) = (0)^4 - 3(0)^3 + 10 = 10$

$\quad f\left(\dfrac{9}{4}\right) = \left(\dfrac{9}{4}\right)^4 - 3\left(\dfrac{9}{4}\right)^3 + 10 = 1.5$

$\therefore$ critical points at: $(0,10), \left(\dfrac{9}{4}, 1.5\right)$

Usually people have no problems in the first three steps of the first derivative test. Finding the critical points, figuring out points of discontinuity, and writing out the intervals are pretty easy.

The problem usually comes in Step #4, where you're asked to find whether the function is increasing or decreasing on the interval. Remember, to do this, you pick some x value within the interval, plug it into f'(x), and see whether the resulting output is (+) or (-)

After making the determination about what the critical points in the interval are, we look at the y-coordinates of all the critical points to see if the endpoints are absolute max's or min's, then we label each point.

As you're probably picking up, finding the critical points is the major step here.

(3) $f'(x)$ exists throughout the real numbers,
     thus no critical points here.

$\therefore$ The critical points for this function are: $(0,10)$, $\left(\dfrac{9}{4},1.5\right)$

[2] $f(x)$ is continuous everywhere

[3] $[-\infty,0]$, $\left[0,\dfrac{9}{4}\right]$, $\left[\dfrac{9}{4},\infty\right]$

[4] $f'(-1) = 4(-1)^3 - 9(-1)^2 = -13$ $\therefore$ $(-)$ $\therefore$ decreasing
     $f'(1) = 4(1)^3 - 9(1)^2 = -5$ $\therefore$ $(-)$ $\therefore$ decreasing
     $f'(3) = 4(3)^3 - 9(3)^2 = 27$ $\therefore$ $(+)$ $\therefore$ increasing

In this example, of special note is the fact that we have a critical point at x=0 that is neither a max nor a min since it has a decreasing interval before and after it.

[5] $x = 0$, decreasing $\rightarrow$ decreasing $\therefore$ not a max or a min

$x = \dfrac{9}{4}$, decreasing $\rightarrow$ increasing $\therefore$ min

$\therefore$ Looking at the critical points then, we see that:

$(0,10)$ is neither a max nor a min

$\left(\dfrac{9}{4},1.5\right)$ is an absolute min

3. $f(x) = 3x^2 + 2x + 1$, $[-3,4]$

[1] Find the critical points.

As usual, we begin by solving for the critical points. If you get this step right, that's half the battle.

(1) $f(-3) = 3(-3)^2 + 2(-3) + 1 = 22$

$f(4) = 3(4)^2 + 2(4) + 1 = 57$

$\therefore$ critical points at: $(-3,22)$, $(4,57)$

$(2)\, f'(x) = 6x + 2$

$$6x - 2 = 0$$

$$x = -\frac{1}{3}$$

$$f\left(-\frac{1}{3}\right) = 3\left(-\frac{1}{3}\right)^2 + 2\left(-\frac{1}{3}\right) + 1 = \frac{2}{3}$$

$\therefore$ critical points at: $\left(-\frac{1}{3}, \frac{2}{3}\right)$

$(3)\, f'(x)$ exits throughout the real numbers,
   thus no critical points here.

$\therefore$ critical points for this function on this interval are:

$(-3,22),\ \left(-\frac{1}{3}, \frac{2}{3}\right),\ (4,57)$

[2] $f(x)$ is continuous everywhere

[3] $\left[-3, -\frac{1}{3}\right], \left[-\frac{1}{3}, 4\right]$

[4] $f'(-1) = 6(-1) + 2 = -4\ \therefore\ (-)\ \therefore$ decreasing
   $f'(0) = 6(0) + 2 = 2\ \therefore\ (+)\ \therefore$ increasing

[5] $x = -\frac{1}{3}$, decreasing $\rightarrow$ increasing $\therefore$ min

$\therefore$ Looking at the critical points then, we see that:

$(-3,22)$ is just an endpoint - -neither an absolute max nor an absolute min

$\left(-\frac{1}{3}, \frac{2}{3}\right)$ is an absolute min

$(4,57)$ is an absolute max

The Second Derivative Test

Having learned the First Derivative Test, we now know how to locate the maximum and minimum points of a function. If we were to plot these points on a graph, we would also know a little bit about the lines that run between them, right? Afterall, if we have a maximum followed by a minimum, we can be fairly confident that the line between the two points connects them in a fairly straight manner. However, when we look at most graphs, lines in-between maximum and minimum points tend to inflect at some point from concave down to concave up. Similarly, lines

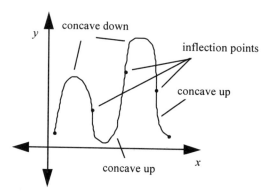

Fig. 4.4: This graph illustrates regions of a graph of a function that are considered to be concave up and concave down. The inflection points occur where the line changes between the two regions.

in-between minimum and maximum points tend to inflect from concave up to concave down. See Fig. 4.4.

Clearly, if we're trying to get an accurate representation of a function, we need more information than just its maximum and minimum points--we need its *inflection points as well*. In order to find these points, we use the Second Derivative Test. Not only does this test enable us to find a function's inflection points, it enables us to quickly determine if a function's critical points are maximums or minimums. Here's the test:

The Second Derivative Test

1. Find the critical points as we did in section 4.1.

2. Find the second derivative of the function.

3. Plug the critical numbers you solved for (x-values of the critical points) into the second derivative. If the result is positive, the critical point is a minimum and the graph in that area is concave up. If the result is negative, the critical point is a maximum and the graph in that area is concave down. Note: Don't plug in the endpoint x-values--it's not necessary.

4. Locate the possible inflection points by setting $f''(x)$ equal to zero and solving for the x's. Inflection points also occur at points where $f''(x)$ is not defined but $f(x)$ is, so visually check for these too. Plug the x's you obtain here into $f(x)$ to get the coordinates of the inflection points. Note: If $f(x)$ is not defined for a given x value, the point is not an inflection point so ignore it.

Examples   4.2.2
These examples are a continuation of those in section 4.2.1, hence we don't need to recalculate the critical points.

1. $f(x) = 3x^3 + 4x^2 - x + 1, \quad [-3,3]$

$[1]$ critical points at: $(-3,-41)$, $(-1,3)$, $\left(\frac{1}{9}, 0.94\right)$, $(3,115)$

$[2]$ $f'(x) = 9x^2 + 8x - 1$
   $f''(x) = 18x + 8$

$[3]$ $f''(-1) = 18(-1) + 8 = -10 \therefore (-)$
   $\therefore$ $(-1,3)$ is a max and the graph is concave down.

$f''\left(\frac{1}{9}\right) = 18\left(\frac{1}{9}\right) + 8 = 10 \therefore +$

$\therefore \left(\frac{1}{9}, 0.94\right)$ is a min and the graph is concave up.

$[4]$ $f''(x) = 18x + 8$
   $18x + 8 = 0$
   $x = -\frac{8}{18} = -\frac{4}{9}$
   $f\left(-\frac{4}{9}\right) = 3\left(-\frac{4}{9}\right)^3 + 4\left(-\frac{4}{9}\right)^2 - \left(-\frac{4}{9}\right) + 1 = 2$
   $\therefore$ inflection point at: $\left(-\frac{4}{9}, 2\right)$

Remember, we already solved for the critical points in 4.2.1, so we don't need to redo that here. Also note, in this problem two of the critical points are due to endpoints, hence we don't use these in step #3. Why? Because we can determine local mins and local maxes can only occur within the interval--and we can visually see whether or not they're absolute maxes or absolute mins just by looking at their y-coordinates.

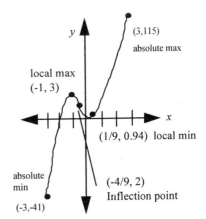

Fig. 4.5 shows the function graphed using the information obtained from the second derivative test. Essentially, all we do is plot the critical points, label them as to what type of point they are, then connect the dots. It's easy--just make sure you label everything.

Fig. 4.5: Note that this graph is not drawn to scale--which is OK since all relevant points are labeled.

2. $f(x) = x^4 - 3x^3 + 10$

[1] critical points at: $(0,10)$, $\left(\dfrac{9}{5}, 1.5\right)$

[2] $f'(x) = 4x^3 - 9x^2$

$\quad f''(x) = 12x^2 - 18x$

[3] $f''(0) = 12(0)^2 - 18(0) = 0$ ∴ $(0,10)$ is a neither a max nor a min.

$f''\left(\dfrac{9}{-}\right) = 12\left(\dfrac{9}{5}\right)^2 - 18\left(\dfrac{9}{5}\right) = 6.5$ ∴ $(+)$ ∴ $\left(\dfrac{9}{5}, 1.5\right)$ is a min and the graph is concave up.

[4] $f''(x) = 12x^2 - 18x$

$\qquad 12x^2 - 18x = 0$

$\qquad x(12x - 18) = 0$

$\qquad x = 0, \; x = \dfrac{18}{12} = \dfrac{3}{2}$

$\quad f(0) = (0)^4 - 3(0)^3 + 10 = 10$

$\quad f\left(\dfrac{3}{2}\right) = \left(\dfrac{3}{2}\right)^4 - 3\left(\dfrac{3}{2}\right)^3 + 10 = 5$

$\quad$ ∴ inflection points at: $(0,10)$, $\left(\dfrac{3}{2}, 5\right)$

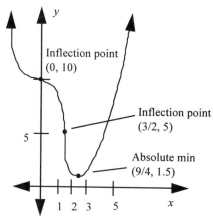

Fig. 4.6

3. $f(x) = 3x^2 + 2x + 1, \; [-3,4]$

[1] critical points at: $(-3,22)$, $\left(-\dfrac{1}{3}, \dfrac{2}{3}\right)$, $(4,57)$

[2] $f'(x) = 6x + 2$

$\quad f''(x) = 6$

[3] $f''\left(-\dfrac{1}{3}\right) = 6$ ∴ $(+)$ ∴ $\left(-\dfrac{1}{3}, \dfrac{2}{3}\right)$

$\quad$ is a min and graph is concave up.

[4] $f''(x) = 6$

$\qquad 6 \neq 0$ ∴ no inflection points.

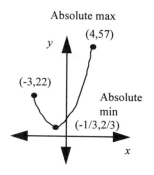

Fig. 4.7

Additional Problems

4.2

Graph the following functions.

1. $f(x) = 7x^2 + 1, \quad [-1, 1]$

4. $f(x) = \dfrac{2x}{x^2 + 5}$

2. $f(x) = 2x^3 - 7x^2 + 4x - 2$

5. $f(x) = x^{\frac{1}{2}} + 2, \quad [0, 4]$

3. $f(x) = -2x^4 + 4x^3 + 1$

4.3 Asymptotes & Limits at Infinity

In the last section we began showing you how to graph functions given only their formula and in some cases an interval. If you look back over the examples, though, you'll notice that we specifically avoided functions that would require the consideration of Step #2 of the First Derivative Test. As you will recall, Step #2 required us to locate any *x*-values that cause the function to be discontinuous or not exist. Conveniently, all of the functions we dealt with were continuous throughout the real numbers, hence we always ended up skipping this step. Unfortunately, there are a lot of functions out there in which certain *x*-values will cause them to be discontinuous or not exist--in other words, their domains are not the real numbers, it's all of the real numbers except a few. For example, the function $f(x) = 1/x$. As you know by now, zero is not in the domain of this function since 1/0 does not exist. Hence, the domain of this function is all the real numbers except zero. If we wanted to graph this function using the First Derivative Test, we'd have to consider Step #2. We're going to spend some time now learning about how to do this.

Another type of function we did not deal with in the previous sections were those whose outputs converge on some number as their inputs move towards infinity--in other words, functions who, as their inputs get greater and greater in either the positve or negative direction, have outputs which approach but never reach a particular number. We're going to spend some time now learning how to identify and graph these types of functions as well. How are we going to do all this? Simple. We're going to use *asymptotes*.

What is an asymptote? Well, before we formally tell you, let's first note that there are two types--vertical and horizontal. *Vertical asymptotes are drawn in places where the inputted x value causes the function to yield an output (y value) equal to some nonzero number divided by zero*, and *horizontal asymptotes occur in places where the output (y value) is the limit of the function as x approaches infinity*.

From these definitions, you can see that vertical asymptotes are the ones that will enable us to graph functions that have points of discontinuity in them, while horizontal asymptotes will help us to more accurately graph functions whose outputs level off and approach, but never reach, a certain value.

Mathematically, these definitions of asymptotes are represented as follows:

Vertical Asymptotes: Denominator of $f(x)$ equals zero, while the numerator
is any number other than zero.

Horizontal Asymptotes: $\lim\limits_{x \to \infty} f(x)$ and / or $\lim\limits_{x \to -\infty} f(x)$

Graphically, we represent asymptotes as straight dashed lines, as in Fig. 4.8.

Now before we get into the solving of problems with asymptotes and then subsequent graphing based on the asymptotes, we need to learn how to solve limits in which x approaches infinity so that we can find horizontal asymptotes. Before we go on then, we'll give you a few simple rules to follow for solving infinite limits, and then run through some examples.

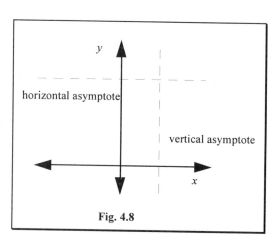

Fig. 4.8

Rules for finding Infinite Limits

1. Plug $x = \infty$ into the function.

(I) If you get some number divided by infinity, then the limit equals zero. (Note: Any number (except infinity) divided by infinity equals zero).

(II) If you get infinity divided by some number, then the limit is infinity. (Note: Infinity divided by any number is infinity).

(III) If you get infinity divided by infinity reduce the expression as before with factoring, or easier yet, concentrate on the highest power variables and ignore the others. For example:

$$\lim_{x \to \infty} \frac{2x^2 + x + 1}{x^2}$$

To solve this note that when infinity is plugged into the function the (+1) on the end is insignificant; it's like saying 5501/10,000 instead of 5500/10,000, there's really no difference so we can ignore it. Likewise, at infinity, the +x is also very tiny compared to x^2, so we ignore it also. This leaves us with just the highest order terms in both the numerator and denominator, which enables us to reduce the and solve. Here we get $2x^2/x^2$, which leads to an answer of, 2.

2. If after solving the limit you find that the limit exists-i.e., the answer is not infinity, then the limit is said to *converge*. If after solving the limit you find that the limit has a + or - infinity as the answer, the limit is said to *diverge*.

Examples   4.3.1

1. $\lim\limits_{x \to \infty} \dfrac{1}{x}$

$\lim\limits_{x \to \infty} \dfrac{1}{x} = \dfrac{1}{\infty} = 0$

Examples #1 and #2 are very straight forward. We just plug in the specified number, infinity, and see what happens. In both cases, as you can see, we get a constant divided by infinity, which results in a limit of zero Hence, the limits both **converge***.*

2. $\lim\limits_{x \to \infty} \dfrac{1}{x^2}$

$\lim\limits_{x \to \infty} \dfrac{1}{x^2} = \dfrac{1}{\infty^2} = 0$

3. $\lim\limits_{x \to \infty} x^2$

$\lim\limits_{x \to \infty} x^2 = \infty^2 = \infty$

Example #3 is similar to the first two in that all we have to do to solve is plug in the specified number, infinity. Here, however, we get infinity squared, which is equal to infinity. Hence the limit **diverges***.*

4. $\lim\limits_{x \to \infty} \dfrac{3x + 1}{x^2 + 2x + 1}$

$\lim\limits_{x \to \infty} \dfrac{3x + 1}{x^2 + 2x + 1} = \lim\limits_{x \to \infty} \dfrac{3x}{x^2}$

$= \lim\limits_{x \to \infty} \dfrac{3}{x} = \dfrac{3}{\infty} = 0$

Example #4 is our first challenging example. We can't just plug infinity in because when we do we get infinity divided by infinity. So, to solve we look for the highest power of x in the numerator, the highest power of x in the denominator, and then trash everything else. We then solve as we normally would. In this case the result is zero--so the limit **converges***.*

5. $\lim\limits_{x \to \infty} \dfrac{2x^2 + 1}{5x^2 + x + 1}$

$\lim\limits_{x \to \infty} \dfrac{2x^2 + 1}{5x^2 + x + 1} = \lim\limits_{x \to \infty} \dfrac{2x^2}{5x^2} = \dfrac{2}{5}$

In this example, we simplify the function quite a bit, and in the process, don't even have to bother plugging in the specified number as we end up with a constant. Therefore the limit **converges**.

6. $\lim\limits_{x \to \infty} \dfrac{x^5 + x^4 + x + 2}{x^3 + 1}$

$\lim\limits_{x \to \infty} \dfrac{x^5 + x^4 + x + 2}{x^3 + 1} = \lim\limits_{x \to \infty} \dfrac{x^5}{x^3}$

$= \lim\limits_{x \to \infty} x^2 = \infty$

In Example #6, we again simplify before plugging in the specified number, though this time the simplifying results in us obtaining a limit of infinity--thus the limit **diverges**.

7. $\lim\limits_{x \to -\infty} \left(-x^2 + x^3\right)$

$\lim\limits_{x \to -\infty} \left(-x^2 + x^3\right) = \lim\limits_{x \to -\infty} \left(x^3\right) = -\infty$

Examples #7 and #8 are the same in that we again are focusing on the highest power terms. The difference between these examples and the previous ones is that the function is not a fraction. In these problems, the biggest dilemma is making sure you keep your signs right so that you know whether you're going to get a + or a - infinity. In both cases, the limtis **diverge**.

8. $\lim\limits_{x \to -\infty} \left(-2x^3 + 4x^2 - 5x + 13\right)$

$\lim\limits_{x \to -\infty} \left(-2x^3 + 4x^2 - 5x + 13\right) = \lim\limits_{x \to -\infty} \left(-2x^3\right)$

$= -2\left(-\infty\right)^3 = \infty$

9. $\lim\limits_{x \to -\infty} \dfrac{10x}{\sqrt{25x^2 - 5x + 2}}$

$\lim\limits_{x \to -\infty} \dfrac{10x}{\sqrt{25x^2 - 5x + 2}} = \lim\limits_{x \to -\infty} \dfrac{10x}{\sqrt{25x^2}} = \lim\limits_{x \to -\infty} \dfrac{10x}{5x}$

$= \lim\limits_{x \to -\infty} \dfrac{10}{5} = 2$

In this last example, the major point of interest is the square root. Note that even with the square root we can simplify as we did in the earlier examples--in this case finding that the limit **converges**.

Finding and Graphing Functions via their Asymptotes

Now that we know how to deal with limits in which x approaches infinity, we can move on to finding asymptotes of functions and using them to construct graphs. To find out whether or not a function has asymptotes, simply look at it and run through the following steps.

Method for determining if a function has asymptotes:

1. Check for vertical asymptotes by noting whether or not the function is a fraction. If it is not, there are no vertical asymptotes. If it is, set the denominator equal to zero and solve for the x-values. The x-values you find are all locations of vertical asymptotes.

2. Check for horizontal asymptotes by taking the limit of the function as x approaches plus or minus infinity. If you get an answer there is a horizontal asymptote at that point, otherwise there are none.

Now that we know how to determine whether or not a function has asymptotes, we can accurately graph any function we run into using the following method:

General Method for Graphing Functions:

1. Determine if the function has asymptotes and draw in those it has.

2. If both vertical and horizontal asymptotes exist, simply plug in a x-value to see which marked off section the function will lie in, and then roughly draw in the function--see examples below.

3. If you have only horizontal or only vertical asymptotes, you have to go back to one of the derivative methods to be really exact. If exactness is not required, you can usually get a fairly accurate picture of the graph by simply plugging in a few points.

4. If there are no asymptotes, use the First and/or Second Derivtive Test to graph.

(NOTE: When drawing the graphs as they approach the asymptotes, make sure they don't touch the asymptotes as the asymptotes mark points where the function does not exist--see the examples below.)

As you can see from the methods outlined above, determining if a function has asymptotes is not too difficult, however, graphing a function with asymptotes can be a bear if it only has one type. Make sure you're thoroughly familiar with the two derivative tests before tackling these types of problems--it can get messy. As long as you follow each step and keep track of where you are, you'll be fine.

Examples   **4.3.2**

1. $f(x) = \dfrac{1}{x+1}$

The first thing we do when we encounter functions that may have asymptotes is check for both vertical and horizontal as is prescribed above. Here we find a vertical asymptote at x = -1, and a horizontal asymptote at y = 0.

[1] vertical asymptote check: $x + 1 = 0$

$x = -1$ yes

Note: Since we have both vertical and horizontal asymptotes, all we need to do to graph is do a sections check.

horizontal asymptote check: $\displaystyle\lim_{x\to\infty}\left(\dfrac{1}{x+1}\right) = 0$ yes

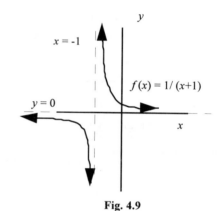

[2] section check: $f(-2) = \dfrac{1}{-2+1} = -1$

$f(0) = \dfrac{1}{0+1} = 1$

Fig. 4.9

2. $f(x) = \dfrac{x^2}{x+1}$

In this example the major complication is that there is a vertical asymptote but not a horizontal, hence we must use one of the derivative tests. We'll use the Second Derivative Test.

[1] vertical asymptote check: $x + 1 = 0$

$x = -1$ $\therefore$ yes

horizontal asymptote check: $\displaystyle\lim_{x\to\infty}\left(\dfrac{x^2}{x+1}\right) = \infty$ $\therefore$ no

[3] Second Derivative Test

 1. Find the critical points.

 (1) No endpoints given, so no critical points here.

 (2) $f'(x) = \dfrac{(x+1)(2x)-(x^2)(1)}{(x+1)^2} = \dfrac{x^2+2x}{(x+1)^2}$

$$\frac{x^2 + 2x}{(x+1)^2} = 0$$

$$x^2 + 2x = 0 \quad \therefore \ x = 0 \ and \ x = -2$$

$$f(0) = \frac{(0)^2}{0+1} = 0$$

$$f(-2) = \frac{(-2)^2}{-2+1} = -4$$

$\therefore$ critical points at: $(0,0), \ (-2,-4)$

(3) $f'(x)$ is discontinuous at $x = -1$, but so is $f(x)$, thus no critical point here.

$\therefore$ The critical points of this function are: $(0,0), \ (-2,-4)$

2. $f''(x) = \dfrac{(x+1)^2(2x+2) - (x^2 + 2x)(2)(x+1)}{(x+1)^4}$

$$= \frac{(x+1)[2x^2 + 4x + 2 - 2x^2 - 4x]}{(x+1)^3} = \frac{2}{(x+1)^3}$$

3. $f''(0) = \dfrac{2}{(0+1)^3} = 2 \ \therefore \ (+) \ \therefore$ min and graph is concave up.

$f''(-2) = \dfrac{2}{(-2+1)^3} = -2 \ \therefore \ (-) \ \therefore$ max and graph is concave down.

$\therefore \ (-2,-4)$ is a local max, and $(0,0)$ is a local min.

4. $f''(x) = \dfrac{2}{(x+1)^3}$

$\dfrac{2}{(x+1)^3} = 0$; No values of x satisfy this equation, $\therefore$ no inflection points.

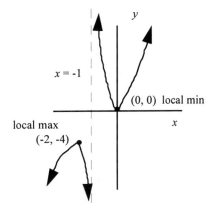

Fig. 4.10

Additional Problems   **4.3**
Find the limits in the following problems.

1. $\lim\limits_{x \to \infty} \dfrac{2}{x}$

9. $\lim\limits_{x \to \infty} \dfrac{7x^2 + 2x + 1}{5}$

2. $\lim\limits_{x \to -\infty} \dfrac{2}{x}$

10. $\lim\limits_{x \to -\infty} \dfrac{-7x^2}{2 + x}$

3. $\lim\limits_{x \to \infty} \dfrac{x + 5}{x}$

11. $\lim\limits_{x \to \infty} \dfrac{\sqrt{x + 1}}{x}$

4. $\lim\limits_{x \to -\infty} \dfrac{x + 5}{x}$

12. $\lim\limits_{x \to \infty} \dfrac{2x^2 + 5x + 1}{\sqrt{x^4 + 2x}}$

5. $\lim\limits_{x \to \infty} \dfrac{2x^2 + 2x + 1}{x^2 - 1}$

13. $\lim\limits_{x \to \infty} \left(3x^2 + 5x + 2\right)$

6. $\lim\limits_{x \to -\infty} \dfrac{5x^3 - 2x + 1}{7x^3 + 2x + 4}$

14. $\lim\limits_{x \to -\infty} \left(2x^3 - 5x + 1\right)$

7. $\lim\limits_{x \to \infty} \left(2x + 1\right)$

15. $\lim\limits_{x \to \infty} \dfrac{7x^5 - 2}{\sqrt{x^2 + 1}}$

8. $\lim\limits_{x \to \infty} \dfrac{x}{2}$

Graph the following functions.

16. $f(x) = \dfrac{7}{x + 6}$

18. $f(x) = \dfrac{x^2}{2x + 5}$

17. $f(x) = \dfrac{8x^4 + 9x^3 + 2}{2x^4 - 4}$

19. $f(x) = \dfrac{9x^2 + 5x - 2}{3x^2 - 12}$

4.4 Story Problems for Max-Mins

By now you're thoroughly familiar with the whole concept of max-mins and the two
derivative tests. Given a function, you can now determine what its maximum and

minimum points are. This is good, because we're about to jump into some very real practical applications of derivatives that require you to do this. What are we talking about? Well, if you remember back to the beginning of this chapter we showed you several Real World Applications and noted that examples in section 4.4 would further illustrate them. So, here we are, ready to do some illustrating.

In this section we're going to be solving max-min story problems--in other words, story problems that require you to find the maximum and minimum values of a function. Usually the most difficult thing about these problems is finding the functions. And, unfortunately, there are no step-by-step guidelines you can follow to guarantee that you'll come up with the correct function. So, what are we going to do? We're going to provide with the following hint, and then expose you to a wide enough variety of story problems that you'll see the most common tricks, and hopefully gain an intuitive sense for how to solve them.

Hint: Always draw a picture of the problem, and write all the given information out in equation form. This will enable you to clarify even the most horrendous problems.

Examples **4.4**

1. In an effort to spread goodwill through the community, your organization decides to fix up one of the old abandoned lots downtown so that the local kids have a safe place to play. Due to the high traffic in the area, it is decided that no matter what is done with the lot, the lot itself should be fenced in. Unfortunately, your organization has only enough money to buy 1,000 feet of good fencing. If you want to surround as much of the lot as possible in a rectangular fashion, what dimensions should the sides be so as to maximize the area within?

[1] Draw out the problem.

Let x = length of the longest side

Let A = area

$A = (\text{length})(\text{width})$

$= x\left(\dfrac{1000 - 2x}{2}\right)$

$= 500x - x^2$

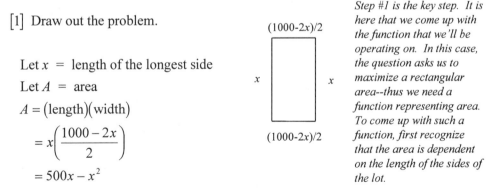

Step #1 is the key step. It is here that we come up with the function that we'll be operating on. In this case, the question asks us to maximize a rectangular area--thus we need a function representing area. To come up with such a function, first recognize that the area is dependent on the length of the sides of the lot.

Hence, we can represent area as a dependent variable, A. Now we all know the equation for the area of a rectangle--length times width--hence we have the makings of a function. To be of use, however, length and width must be given in terms of just one variable. In order to do this, we have to consider the given information--namely that the lot is rectangular and that its perimeter can be no more than 1000 feet. Since it's a rectangle, the opposite sides must be the same. If we let one of the sides equal x, its the opposite side must also equal x. Additionally, since there's only 1000 feet of fence, there's only 1000 - 2x left for the other two sides. Since both those sides must be equal to each other in length, they are each (1000 - 2x)/2 long. x then is the independent variable.

[2] Fid the max area

function: $A = f(x) = 500x - x^2$

Once we have the function, all we have to do is find its maximum value. To do this we first find its critical numbers, which, since no interval is given simply involves taking the first derivative, setting it equal to zero and solving.

$$\therefore \ \frac{dA}{dx} = f'(x) = 500 - 2x$$

$$500 - 2x = 0$$

$$x = 250 \ \Leftarrow \text{critical number}$$

Note: Don't let this dA/dx notation throw you. As you can see, it's equivalent to f'(x). Think back to some of the alternative notations in Chapter 3--remember dy/dx. That's all this is here. It's just that we've got A as our dependent variable instead of y.

[3] Check to make sure this value is a

max point via the second derivative test:

$$\frac{d^2 A}{dx^2} = f''(x) = -2$$

$$f''(250) = -2 \ \therefore \ (-) \ \therefore \ \text{a max}$$

Upon finding the critical number, we check to see if it's a max using a portion of the second derivative test.

After confirming that it is in fact a max, we just answer the question.

2. Our second Real World Application in this chapter dealt with an engineer having to design a battery case using as little material as possible while containing some specified volume. Well, suppose the specifics of the problem are as follows: The desired volume to be encompassed is 1000 cubic inches and the case is to have a square base with an open top. Solve.

[1] Let h = height

Let x = length of base

Let V = volume

Let A = surface area

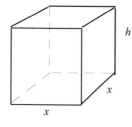

Same idea as the last problem except this time we're trying to minimize a function. Again, the key step is coming up with the function. Here, the function we're trying to minimize is surface area. Hence, A will again be our dependent variable.

$$V = x^2 h$$

$$A = x^2 + 4xh$$

Unfortunately, the equation for surface area has two independent variables--x and h. Hence, we need to use another equation to eliminate one of these variables. How do we do this? Well, since we're given the volume, we can solve the volume equation for either h or x and then make a simple substitution into the area equation--thus reducing the surface area equation's independent variables down to one. This is shown on the next page.

[2] $x^2 h = 1000$

$$h = \frac{1000}{x^2}$$

$$A = x^2 + 4x\left(\frac{1000}{x^2}\right)$$

$$A = x^2 + \frac{4000}{x}$$

Due to the extra manipulation, we need a second step in this problem. Hence, here in step #2 we solve the volume equation to get h in terms of x. Then we go to the area equation and substitute this new found value for h in. Result: The area equation now has just one independent variable. Remember, y is the dependent variable.

[3] $\dfrac{dA}{dx} = 2x - \dfrac{4000}{x^2}$

$$2x - \frac{4000}{x^2} = 0$$

$$2x = \frac{4000}{x^2}$$

$$2x^3 = 4000$$

$$x = (2000)^{\frac{1}{3}} = 12.6$$

$$\therefore \ h = \frac{1000}{(12.6)^2} = 6.3$$

Step #3 we find the minimum values of the function by first finding its critical numbers.

Again, don't let the dA/dx notation throw you--it's the same as saying f′(x), or dy/dx if our dependent variable is named y instead of A.

[4] Make sure this is a min and not max.

$$\frac{d^2 A}{dx^2} = 2 + \frac{8000}{x^3} \quad \therefore \ (+) \ \therefore \ \text{concave up} \ \therefore \ \text{it's a min}$$

$\therefore$ dimensions of the box for minimu use of materials: base = 12.6 in.

height = 6.3 in.

Note: If the engineer had been in a foul mood while designing the case and decided that instead of calculating the minimum amount of material to use, he was going to calculate the maximum, how would he have done this?

(I) Realize that since the only critical number found was a min, the max must come at one of the endpoints. Assuming that the smallest value x or h can take on is 1 inch, these values are simply plugged into the volume and area functions to determine which combination will yield the most surface area.

Let $h = 1$ inch, find x:

$x^2(1) = 1000$

$\quad x = 31.6 \quad \therefore \quad (31.6,\ 1)$

Let $x = 1$ inch

$(1)^2 h = 1000$

$\quad h = 1000 \quad \therefore \quad (1,\ 1000)$

$A = (31.6)^2 + \dfrac{4000}{31.6} = 1125\ \text{in}^2$

$A = (1)^2 + \dfrac{4000}{1} = 40001\ \text{in}^2$

$\therefore$ To use the most material the dimensions would be: base $= 1$ in.

height $= 1000$ in.

3. As you will recall, the first Real World Application we talked about in this chapter had to do with maximizing a company's profits. Here's an example of how this is done: Suppose that *SPR Inc.*, a manufacturer of sporting equipment, charges $100.00 for a pair of knee braces. If the cost of manufacturing each pair of knee braces is given by the function:

$$C = 1.1x^2 + 9x + 10, \text{ where } x = \text{ the number of pairs of knee braces}$$

figure out how many pairs of knee braces should be produced each day to maximize profits (assume that *SPR* can sell everything it makes--these braces are in high demand--and that, at most, 50 pairs of braces can be produced).

[1] Let $P = $ profit $= $ income - expense

$P = 100x - \left[1.1x^2 + 9x + 10\right]$

$P = 91x - 1.1x^2 - 10$

Again, the key is getting the original function. Here, we simply define profit, then plug in our given information: We make $100 per pair of knee brace--i.e., income, 100x, and our expense--i.e., our cost is given.

[2] $\dfrac{dP}{dx} = 91 - 2.2x$

$\quad 91 - 2.2x = 0$

$\quad\quad x = 41.4 \approx 41$

As before, we take the derivative of this function in an effort to locate the critical numbers. Then we check to see, in this case, if the found number is a max. using a bit of the Second Derivative Test.

[3] Check to see if x is a max:

$\dfrac{d^2 P}{dx^2} = -2.2 \quad \therefore \quad (\text{-}) \quad \therefore \quad \text{a max}$

[4] Check end points to see if x is the absolute max (0 braces, 50 braces).

$$P(0) = 100(0) - \left[1.1(0)^2 + 9(0) + 10\right] = -10$$

$$P(50) = 100(50) - \left[1.1(50)^2 + 9(50) + 10\right] = 1,790.00$$

$$P(41) = 100(41) - \left[1.1(41)^2 + 9(41) + 10\right] = 1,871.90$$

Therefore, manufacturing 41 pairs of knee braces a day will maximize profits.

Additional Problems   **4.4**

Solve the following story problems.

1. Suppose you're working for a home decorating company and they come to you with the following problem: The company has just received a shipment of ten thousand 10-inch pieces of wire which they intend to use as the base structure for a new line of decorative baskets. What they would like you to figure out is how to cut the wire so as to make a circular base and an equilateral triangle base so that the sum of the two areas is (1) minimum, and (2) maximum.

2. Close friends of yours have decided to build a storage garage that will enclose a volume of 70,000 cubic feet. Obviously, they don't want to spend any more money than they have too, so they ask you what dimensions they should use in construction so as to minimize the amount of necessary material. The only constraints they put on your design is that the garage should have a base in the shape of a square. The material they are concerned about minimizing is for the sides and roof of the garage, and due to the fact that the garage will be bordering a barn, only three walls are needed. Specify the dimensions.

3. You've decided to open a *Bed 'n Breakfast* in Maine. After careful planning and several market surveys, you figure you can charge anywhere from $30.00 to $100.00 per person for a room. A consultant you've hired to help you get the business up and running tells you that, based on past experiences, you can assume that the number of people renting rooms will depend on the price you charge, x, as $x - 0.008x^2$. If you want to maximize your profits, how much do you charge?

4. How about that! Your best friend just won the *Elox Sweepstakes*. *Elox Company* promises to give her a steel cone (base radius 3 inches, height 6 inches), and tells her that they will stick a solid gold cylinder in it in such a way that the axis of the cone and the cylinder coincide. She has to decide the dimensions of the cylinder, though. (Note: The cylinder can't stick out above the cone). After a few moments reflection, she asks you for some help.

4.5 Other Derivative Story Problems: Rates and Related Rates

Guess what? We're finally finished with max-mins! We're moving on to the last practical application we're going to be seeing in this chapter--story problems involving rates and related rates. Compared to max-mins, these problems are easy. You see, rate problems work off of the basic premise that given a function that relates distance, s, as a function of time, t, the first derivative of the function will give you the velocity as a function of time, and the second derivative will give you acceleration as a function of time. Mathematically this is represented as:

$$\text{Given } s = f(t), \text{ then}$$

$$f'(t) = v = \frac{ds}{dt}$$

$$f''(t) = a = \frac{dv}{dt} = \frac{d^2 s}{dt^2}$$

A common equation employed in many rate questions (which you'll see a lot on exams) is the *free fall* equation:

$$s = s_o + v_o t - 16t^2$$

--where s is the total distance traveled in time t, so is the initial height of the falling object, v_o is its initial velocity, -16 is the acceleration due to gravity divided by 2, and t is its flight time.

Related rate problems, on the other hand, are a little more abstract--they don't have any set equations to go by. We're usually given a problem with two rates and must figure out how to find one given the other. Usually these problems are just a matter of setting up the correct equations and then plugging and chugging. As long as you remember that the given rates will be represented by derivatives, you should have an easy time of it.

(Note: We've been working really hard lately, so we're going to end this chapter with some not-so-serious story problems. We hope you find these to be relaxing.)

Examples: Rates 4.5.1

1. You're standing on top of a 125 ft. cliff, and just for laughs, begin rolling boulders over the side. How long does it take a boulder to hit the ground? What is its velocity on impact? Its acceleration on impact?

Use the free fall equation, $s = s_o + v_o t - 16t^2$

Given: $s_o = 125$

$v_o = 0$

Time to impact:

$s = 0$ when the boulder hits the ground

$\therefore \ 0 = 125 + (0)t - 16t^2$

$125 = 16t^2$

$t = 2.8 \quad \therefore \ t = 2.8$ seconds

Velocity on impact:

$\dfrac{ds}{dt} = -32t = -32(2.8) = -89.6$

$\therefore \ v = -89.6 \ \dfrac{\text{ft}}{\text{sec}}$

Acceleration on impact:

$\dfrac{dv}{dt} = \dfrac{d^2 s}{dt^2} = -32$

$\therefore \ a = -32 \ \dfrac{\text{ft}}{\text{sec}^2}$

In order to solve this problem you have to start by getting the given information translated correctly. In the free fall equation, the easiest way to do this is to think of distance s, as being a measure of height.

Thus, when we're told our initial starting position is 125 ft up, we know s_o is 125. When we're told our final position is the ground, a height of 0, we know s at that time is 0. Also, since we're dropping things, not throwing them, they don't start with any initial velocity. Hence, v_o is 0.

2. After rolling boulders for a while, you pull out your sling shot and start launching stones straight up into the air. If the initial velocity is 100 ft/sec, what is the stone's maximum height, how long does it take to reach this maximum height, how long does it take to fall back to the ground from this maximum height, and how fast is it going when it hits? Also, what is the total round trip time?

Begin by realizing that you need to use the free fall equation, $s = s_o + v_o t - 16t^2$

[1] Time to max height

Given: $v_o = 100$, $s_o = 0$

$\therefore\ s = 100t - 16t^2$

$\dfrac{ds}{dt} = 100 - 32t$,   at max height $v = 0$, so

$\quad 0 = 100 - 32t$

$\quad 3.1 = t$, $\therefore\ t = 3.1$ seconds

[2] Max height

$s = 100t - 16t^2$

$\quad = 100(3.1) - 16(3.1)^2$

$\quad = 156.2$ ft

[3] Time to fall back to earth

$v_o = 0$, $s_o = 156.2$, $s = 0$

$\quad 0 = 156.2 - 16t^2$

$\quad 156.2 = 16t^2$

$\quad 3.1 = t$, $\therefore$ Round trip time $= 6.2$ sec.

[4] Velocity on impact

$s = 156.2 - 16t^2$

$\dfrac{ds}{dt} = -32t = -32(3.1) = -99.2$

$\quad v = -99.2\ \dfrac{\text{ft}}{\text{sec}}$

3. Unfortunately, you neglected to move aside after launching the stone and thus got clocked pretty good. As you're being carried on stretcher to the waiting ambulance, you discover the route to the hospital is given by: $s = t^3 - 3t^2 - 9t$. Describe the motion of the ambulance through this route--it takes 10 minutes to get to the hospital. The interval is thus [0, 10].

[1] $v = \dfrac{ds}{dt} = 3t^2 - 6t - 9$

$\qquad = (3t - 9)(t + 1)$

[2] velocity $= 0$ at $3t - 9 = 0$

$\therefore\ 3t = 9$

$t = 3$

velocity $= 0$ at $t + 1 = 0$

$\therefore\ t = -1\ \Leftarrow$ out of interval so ignore

[3] Divide interval based on the pertinant critical numbers:

$(0,\ 3),\ (3,\ 10)$

$f'(1) = 3(1)^2 - 6(1) - 9 = -12\ \therefore\ (\text{-})\ \therefore$ left

$f'(4) = 3(4)^2 - 6(4) - 9 = 15\ \therefore\ (\text{+})\ \therefore$ right

[4] Determine the location at the critical points:

$f'(0) = 3(0)^2 - 6(0) - 9(0) = 0$

$f'(3) = 3(3)^2 - 6(3) - 9(3) = -27$

$f'(10) = 3(10)^2 - 6(10) - 9(10) = 610$

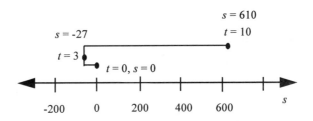

So, the ambulance travels left for 3 seconds
and right for the remaining 7 seconds.

Additional Problems: Rates   4.5.1
Solve the following story problems.

1. As you're lying in your hospital bed recovering from your head wound, you turn the TV on and see a satellite being launched into orbit. It takes 3 seconds for the satellite to travel the first 1500 feet of its journey. What will be its maximum height, and how long will it take to get there?

2. After doing these calculations you realize the satellite won't get high enough to get into orbit. How long does it take to crash after reaching maximum height, what was the round trip time, and how fast does it hit? Also, what is its acceleration when it hits?

3. Realizing their error 3 seconds into the satellite's flight, the scientists on the ground jump into a truck and take off. Their frantic driving produces a route given by:

$$s = t^3 - 35t^2 + 200t$$

The interval of interest is [3, 34.2]. Diagram their route.

Examples: Related Rates   4.5.2

1. It's one of those days when you're just feeling ornery, so you decide to pull a prank on one of your friends. Since you don't want to be noticed, you wait until after dark and then proceed as inconspicuously as possible with your 50 foot ladder to his dorm (he lives on the third floor--window 35 feet up). To get the ladder up to the window, you start by leaning one end of it against the building 10 feet up. You then go to the other end and start moving with the ladder towards the building at a speed of 2 ft./sec. How fast is the end against the building moving up when it is at a height of 25 ft.?

[1] Rate #1: Your speed moving towards the building

   Rate #2: The other end of the ladder's speed moving up

   Relationship: A triangle

let y = distance between ladder and ground

let x = distance between building and you

$\therefore$ $\dfrac{dy}{dt}$ = velocity of the ladder going up

$\dfrac{dx}{dt}$ = velocity of you moving towards the building

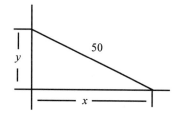

[2] The sides of the triangle are related to each other via the

   Pythagarean Theorem: $x^2 + y^2 = 50^2$

$$\frac{d}{dt}\left[x^2 + y^2 = 50^2\right]$$

$$2x\frac{dx}{dt} + 2y\frac{dy}{dt} = 0$$

When $y = 25$, $x = ?$

$$x^2 = (50)^2 - (25)^2$$
$$x = 43.3$$
$$\therefore \ 2(43.3)(-2) + 2(25)\frac{dy}{dt} = 0$$
$$\frac{dy}{dt} = 3.5 \ \frac{ft}{sec}$$

2. Having successfully gotten the ladder in place, you proceed up with your assortment of shaving cream, balloons, etc. Unfortunately, just as you get to the windowsill you feel the ladder begin to slide down the side of the building. As you cling to the windowsill with both hands for dear life, you look down and see your friend pulling the bottom of the ladder away from the building at a constant velocity of 10 ft./sec. How fast is the ladder moving down when it is 10 ft. above the ground.

[1] Rate #1 = Velocity of friend

Rate #2 = Velocity of falling ladder

Relationship = Triangle

Look back to the figure in the previous example to get a "picture" of what's going on here.

[2] Let x = distance between building and friend

Let y = distance between ground and ladder

$$\therefore \ \frac{dy}{dt} = ? \qquad \frac{dx}{dt} = +10 \ \frac{ft}{sec}$$

As in the last example, the sides of the triangle are related to each other via the Pythagorean Theorem. To get the rates then, we just take the derivative of it:

$$\frac{d}{dt}\left[x^2 + y^2 = 50^2\right]$$
$$2x\frac{dx}{dt} + 2y\frac{dy}{dt} = 0$$

Solving for, and then plugging in the relevant values:

$$at \ y = 10$$
$$x^2 = (50)^2 - (10)^2$$
$$x = 49$$
$$\therefore \quad 2(49)(10) + 2(10)\frac{dy}{dt} = 0$$
$$\frac{dy}{dt} = -49 \ \frac{ft}{sec}$$

3. Naturally you become indignant at your friend's total disregard for your health. So you haul yourself up through the window and head straight to the water faucet. As you're filling up one of your water balloons to launch at him while he rolls around on the ground laughing, you notice that your balloon is taking the form of a perfect cone, height of 10 inches, base radius of 4 inches. Calculate how fast the water is rising in the cone when the water is at a height of 8 inches, if the water from the faucet is flowing in at a rate of 4 cubic inches per second.

[1] Rate #1 = Rate water is flowing in

    Rate #2 = Rate height is increasing in cone

    Rate #3 = Rate the radius is increasing

    Relationship = Volume equation for a cone: $V = \dfrac{1}{3}\pi r^2 h$

[2] $V = \dfrac{1}{3}\pi r^2 h$

We need to get r in terms of h before doing anything here or when we

differentiate, we'll have 2 unknowns $\left(\dfrac{dr}{dt} \text{ and } \dfrac{dh}{dt}\right)$ and only 1 equation.

$\therefore$ when $r = 4$, $h = 10$

$$\dfrac{r}{h} = \dfrac{4}{10}$$

$$10r = 4h$$

$$r = \dfrac{2}{5}h, \qquad \therefore V = \dfrac{1}{3}\pi\left(\dfrac{2}{5}h\right)^2 h = \dfrac{4}{75}\pi h^3$$

Now that we have our volume equation in terms of just one variable - -the variable of interest, h - -we simply have to differentiate it to figure out how fast h is changing.

[3] $\dfrac{dV}{dt} = \dfrac{12}{75}\pi h^2 \dfrac{dh}{dt}$

$$4 = \dfrac{12}{75}\pi (8)^2 \dfrac{dh}{dt}$$

$$0.12 = \dfrac{dh}{dt}$$

$$\therefore \quad \dfrac{dh}{dt} = 0.12 \ \dfrac{\text{in}}{\text{sec}}$$

4. As you come to the window to launch the balloon, your friend sees you and takes off running at a constant velocity--big mistake, big. If he is running at 8 feet per second away from the building, how fast is he moving away from you when he's 20 feet from the building?

[1] Rate #1 = Rate he's moving away from the building

  Rate #2 = Rate the straight line distance between you

     and him is changing

  Relationship = Triangle

[2] Let y = straight line distance between you and him

  Let x = distance between him and the building

Given: $x = 20$, we can use the Pythagorean Theorem to solve for y:

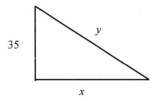

$$y^2 = x^2 + (35)^2$$

$$y = \sqrt{(20)^2 + (35)^2} = 40.3$$

$$\frac{d}{dt}\left[x^2 + (35)^2 = y^2\right]$$

$$2x\frac{dx}{dt} = 2y\frac{dy}{dt}$$

$$\therefore \ 2(20)(8) = 2(40.3)\frac{dy}{dt}$$

$$3.97 = \frac{dy}{dt}$$

$$\therefore \ \frac{dy}{dt} = 3.97 \ \frac{\text{ft}}{\text{sec}}$$

Additional Problems   **4.5.2**
Solve the following story problems

1. After successfully launching and hitting your friend with the water balloon, you decide to quickly fill another before he finds you. As you're filling this next balloon you note that this time it is taking on a spherical shape. If the water is still coming in at 4 cubic inches per second, how fast is the radius changing when the volume is 200 cubic inches?

2. Just as you finish filling the balloon your friend comes flying into the room in a blind panic. It seems the person you saw running away from the building that you hit with the balloon wasn't your friend--he was gone putting the ladder away--it was the RD. Since the RD is a she, it will not do at all to explain you mistook her for your male friend. Clearly you must make a run for it.

You take off down the hall and make a quick left. Just as you do this, she gets to the top of the stairs and comes running down the hall after you. Fortunately she neglects to turn left and keeps right on going. If she's moving at 8 feet per second, and you're moving at 15 feet per second, how fast are you moving apart 10 seconds after she misses the turn, assuming you were 25 feet away when she got to the turn?

3. Having escaped the angry clutches of the RD, you start making your way back to your apartment. As you're walking along, you pass under a street light. If the light is 20 feet up, you're 5 feet 6 inches tall, and walking at a rate of 3 feet per second, how fast is the length of your shadow increasing when you're 19 feet away from the base of the pole?

Integrals

In this chapter we begin our study of the third and final mathematical operation that comprises calculus--the integral. As you are about to see, integrals are closely related to derivatives. Specifically, integrals are defined via limits much as derivatives are, and integrals mathematically reverse what derivatives do. Pretty cool, eh? While there's a lot for us to learn about integrals--it's going to take us the next four chapters to get through it all--for the most part they are pretty straight forward. Simply put: You won't find this branch of calculus to be too difficult .

De'ja vu. Ever experience it before? Well, if you haven't you're about too. Why? Because you just finished learning about derivatives and are now going to learn about integrals. End result--you're going to think you've see things before that you actually haven't. Why? Because integrals reverse what derivatives do. You'll soon see what is meant by this. Before you do, however, we have to get some administrative stuff out of the way. Namely, we have to point out that our study of integrals is not going to be a quick one or two chapter and out deal like limits and derivatives were. You see, integrals are a bit more involved. Not only do we have to learn about their fundamentals and applications, we have to learn about "funky" functions that they often operate on as well as advanced techniques for solving some of the nastier ones. Hence, we'll be spending the next four chapters on integrals. Fortunately, we're going to get off to a good start as this chapter is basically a walk in the park.

5.1 The Basis of the Integral

With four chapters on the same topic ahead of us, it's best that we get right to it. So, we begin this chapter immediately with the following definition:

An integral is a mathematical operation that determines the net area between the graph of a function and a chosen axis over some interval on that axis.

The emphasis here is on the words *net area*. Integrals are net area determiners, that is all. Here's how they work:

Suppose we want to find the area between the function $f(x) = x^2 + 1$ and the x-axis, over the interval $x = 1$ to $x = 4$.

The integral begins by dividing the part of the x-axis between 1 and 4 into equal intervals. It then draws rectangles in these intervals with heights varying so as to reach the function. See Fig. 5.1.

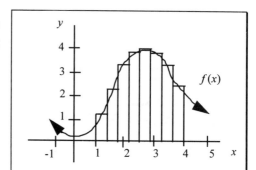

Fig. 5.1: A rough estimate of the area between the function, $f(x)$, and the x-axis over the interval $x = 1$ to $x = 4$, using a few rectangles to estimate.

Now imagine, if the integral stopped and calculated the areas of each of these rectangles and then added them all together-- it would have a pretty good approximation of the desired area. However, if the integral decreased the lengths of the intervals, say enough to double the number of rectangles, the approximation would be better, right? In fact, if the integral was able to construct intervals infinitely close to one another, so as to get an infinite number of rectangles, the approximation of the area would be superb. This is exactly what the integral does. See Fig. 5.2.

As you have probably already surmised, the task of constructing the intervals infinitely close to one another involves limits, while the task of adding the infinite number of rectangles involves summations. Mathematically, we represent all this as follows:

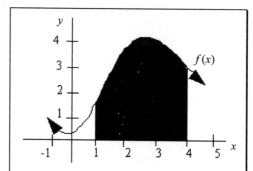

Fig. 5.2: With an infinite number of intervals, all you can see between the function and the axis are the sides of the rectangles. The result is a perfect approximation of the area.

The area between the function, $f(x)$, and the x - axis: $A = \lim\limits_{\Delta x \to 0} \sum\limits_{i=1}^{n} f(t_i)\Delta x_i$

--where Δx_i = the length of the intervals,

$f(t_i)$ = the heights of the various rectangles.

To eliminate the limit and the summation, we use a new notation:

Notation:

$$\lim_{\Delta x \to 0} \sum_{i=1}^{n} f(t_i) \Delta x_i = \int_a^b f(x)dx$$

--which is read: the integral of the function *f* of *x* from *a* to *b*.

--where $\int$ is the integral sign, and *a* and *b* are the endpoints of the interval.

Now, there is a slight complication to all of this which we haven't yet addressed. If you look back at the definition of the integral which we provided on page 95, you'll note that we said integrals find *net area*. What do we mean by this? Well, if we look at the integral notation above we see that, mathematically, what we're doing is calculating the areas of a lot of rectangles and adding the results together. As you will recall, the formula for the area of a rectangle is length times width. In the above notation, $f(x)$ represents the length and *dx* the width.

This being the case, what happens when the graph of the function falls below the *x*-axis-- what happens to the value of $f(x)$? That's right, it becomes negative. Hence, when we multiply it by the positive *dx*, we get a negative area. Thus if we have a function that runs above and below the *x*-axis and we take the integral of the function over an interval in which this occurs, the answer that the integral provides is the net area--the positive area subtracted from the negative area. See Fig. 5.3.

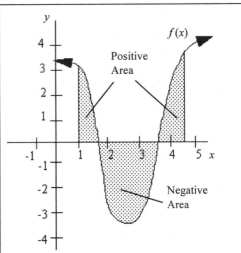

Fig. 5.3: The graph of a function, $f(x)$, which runs above and below the *x*-axis over the interval $x = 1$ to $x = 4.5$. Note that in these types of cases we get both positve and negative area. The negative area comes from the negative values of $f(x)$--example, at $x = 2$, $f(x) = -2.9$.

5.2 Antiderivatives

Now that we know what integrals are and what they're used for, let's learn how to evaluate them. The key to all integral evaluations is finding the *antiderivative* of the function that is being integrated. What is an antiderivative? A quick example will answer this question best. (We're getting into the de'ja vu part).

Suppose we take the derivative of the function $f(x) = 5x$, then $f'(x) = 5$. We say that $f'(x)$ is the derivative of $f(x)$, and that $f(x)$ is the *antiderivative* of $f'(x)$. In other words, you essentially work backwards when asked to find an antiderivative. You must say to yourself, *"What function, by taking its derivative, would give me the function that I'm looking at right now?"*

Admittedly this is kind of tricky. Fortunately, at this point we're not going to attempt to learn all the rules for every type of function out there in one sitting. Rather, we're going to start learning these rules piece by piece, one type of function at a time. We begin with the simplest case:

Rules for finding the antiderivatives of functions in the form, $f(x) = ax^n$.

Given a function $f(x) = ax^n$, and asked to find the antiderivative, $F(x)$,

$$F(x) = \frac{ax^{(n+1)}}{n+1} + C; \text{ where } C \text{ is a constant.}$$

Note: The only sure way to understand this is to work lots of examples; it's just a matter of mental gymnastics.

Examples   5.2

1. $f(x) = x^2$

$$F(x) = \frac{1}{(2+1)} x^{2+1} = \frac{1}{3} x^3 + C$$

At first glance this looks really nasty, but it's not. All we're doing is figuring out what function has its first derivative equal to f(x). The way we do that is to add one to the exponent, then take the reciprocal of the new exponent and multiply the entire function by it.

In this case the exponent is 2, so we add 1 to it, getting 3 for our new exponent, then we just multiply the whole thing by 1/3. To verify that we're correct, we simply take the derivative of F(x)--if we did it write we'll get f(x).

2. $f(x) = 3x^2$

$$F(x) = \frac{3}{(2+1)} x^{2+1} = \frac{3}{3} x^3 = x^3 + C$$

Another example just like the previous one, the only difference being that we have the constant 3 included in f(x) this time. The process for finding the antiderivative is exactly the same, the constant in front has no bearing on the mechanics of cranking out the antiderivative.

Remember, if we did it right the derivative of F(x) will equal f(x).

3. $f(x) = -2x^4$

$$F(x) = \frac{-2}{(4+1)} x^{4+1} = -\frac{2}{5} x^5 + C$$

Just a slight variation on what we've seen so far, no big deal.

4. $f(x) = 3x^{-2}$

$$F(x) = \frac{3}{(-2+1)} x^{-2+1}$$

$$= \frac{3}{-1} x^{-1} = -3x^{-1} + C$$

The only difference between this examples and the others is that the original exponent is a negative number. Hence, when we add a positive 1 to it, it becomes less negative. Obvious, yes. But you'd be surprised how many folks flail on this the first time they see it--we don't want you to be one of them.

5. $f(x) = 7x^2 + 3x^3 + x^{-2} + 5$

$$F(x) = \frac{7}{(2+1)} x^{2+1} + \frac{3}{(3+1)} x^{3+1}$$

$$+ \frac{1}{(-2+1)} x^{-2+1} + \frac{5}{(0+1)} x^{0+1}$$

$$= \frac{7}{3} x^3 + \frac{3}{4} x^4 - x^{-1} + 5x + C$$

This example is filled with good information. First of all, note that when we have a function with several different terms, we treat each one separately.

Second, note that when we have a constant and are seeking its antiderivative, we end up getting a variable--the steps are worked out completely for you to see.

6. $f(x) = \cos x$

$$F(x) = \sin x + C$$

Simple example to show you that we do in fact find antiderivatives of functions that don't fit the standard pattern we've been showing you so far. In this case, it's very simple. To verify that we've done it right, again, just take the derivative of F(x)--it better equal f(x).

7. $f(x) = \sin 8x$

$$F(x) = -\frac{1}{8} \cos 8x + C$$

This example is not nearly as easy to solve as the last one. Not only do we not have a pattern to follow, it's a fairly complex function. Basically, the only way to solve this is to recognize what function, upon taking its derivative, would give you the function you're looking at.

It helps to do this in steps--example: we have a sine in f(x), thus you know there will be a cosine in F(x). Also, since there's no (-) in f(x), you know F(x) must have one since the derivative of cosine is -sine. If you work back like this, step-by-step, you'll eventually figure it out. Remember, to check, take the derivative of F(x)--you'll get f(x) if you did it right.

8. $f(x) = (x+3)^2$

$$F(x) = \frac{1}{2+1} (x+3)^{2+1} = \frac{1}{3} (x+3)^3 + C$$

Another nontrivial example. Don't get real stressed out if you have difficulty finding the antiderivatives of these last few examples. There are advanced methods for dealing with these sorts of problems. The only reason you're introduced to them here is, as far as we can tell, to expand your mind.

9. $f(x) = \sqrt[5]{243x^3}$

$f(x) = 243x^{\frac{3}{5}}$

$F(x) = \dfrac{243}{\left(\dfrac{3}{5}+1\right)} x^{\frac{3}{5}+1} = \dfrac{243}{\left(\dfrac{8}{5}\right)} x^{\frac{8}{5}} = \dfrac{1215}{8} x^{\frac{8}{5}} + C$

This example is essentially an exercise in determining how to express radicals. If you see the trick once, you'll never be fooled again. Watch out for it. This type of problem has been known to appear on exams on more than one occasion.

10. $f(x) = \dfrac{10x^2 + 5x^2 \cos x + 4}{x^2}$

$f(x) = 10 + 5\cos x + 4x^{-2}$

$F(x) = 10x + 5\sin x - 4x^{-1} + C$

Usually, when people see this question they panic, thinking they're going to have to come up with some really nasty antiderivative. In fact, all you have to do is simplify the function first, then go to work. As you can see, after simplifying, the problem becomes trivial.

Additional Problems   5.2
Find the antiderivatives of the given functions.

1. $f(x) = 2x$

2. $f(x) = x$

3. $f(x) = 2x^3$

4. $f(x) = -18x^2$

5. $f(x) = -19x^3$

6. $f(x) = -96x$

7. $f(x) = x^{-2}$

8. $f(x) = x^{-3}$

9. $f(x) = 2x^{19}$

10. $f(x) = 10x^{-2}$

11. $f(x) = \cos 2x$

12. $f(x) = \sin 19x$

13. $f(x) = 9x^2 + 8x - 2$

14. $f(x) = -x^4 - 4x^3 - 10x^2 - 1$

15. $f(x) = \cos x + 27x^2 - 11x + 2$

16. $f(x) = \sin^2 x - 1996 + \cos^2 x$

17. $f(x) = (x-2)^2$

18. $f(x) = (x-2)(x+5)$

19. $f(x) = (x+10)(x+2)$

20. $f(x) = \sqrt[7]{128x^2}$

21. $f(x) = \dfrac{10x^2 + x^2 \sin 2x + 19}{x^2}$

22. $f(x) = \dfrac{110x^4 + 7x^2 + 51}{x^4}$

5.3 Evaluating Definite Integrals

Now that we know the basic idea behind how integrals operate as well as how to compute antiderivatives, it's time we learn how to evaluate integrals. Before we do, however, we need to define *definite* and *indefinite integrals*.

- A *definite integral* is an integral with boundary points specified--in other words, interval is specified.

- An *indefinite integral* is an integral in which no boundary points are given--in other words, the interval is not specified.

These designations are significant in that they require different approaches in solving. We begin with the definite case:

Method for solving definite integrals:

1. Find the antiderivative of the function within the integral.

2. Plug the two boundary points into the antiderivative and subtract the resulting expressions from each other.

$$\int_a^b f(x)dx = F(x)\Big|_a^b = F(b) - F(a)$$

(We provide a list of rules below to aid in this process).

Helpful Rules for Evaluating Integrals

1. $\int_a^b f(x)dx = F(x)\Big|_a^b = F(b) - F(a)$

2. $\int_a^b k[f(x)]dx = k\int_a^b f(x)dx;$ where k is a constant

3. $\int_a^b [f(x) + g(x)]dx = \int_a^b f(x)dx + \int_a^b g(x)dx$

4. $\int_a^b [f(x) - g(x)]dx = \int_a^b f(x)dx - \int_a^b g(x)dx$

5. $\int_a^b f(x)dx = \int_a^c f(x)dx + \int_c^b f(x)dx;$ where $a < c < b$

Examples   5.3

1. $\int_{-1}^{1} x^2 \, dx$

Very simple example to start off with. All you do is take the antiderivative of the function f(x), (f(x) is x^2 in this case), obtaining 1/3 x^3, and then you just follow the guidelines outlined in Rule #1 above.

$$\int_{-1}^{1} x^2 \, dx = \frac{1}{3} x^3 \Big|_{-1}^{1} = \frac{1}{3}(1)^3 - \left(\frac{1}{3}\right)(-1)^3$$

$$= \frac{2}{3}$$

2. $\int_{-2}^{2} 2x^3 \, dx$

$$\int_{-2}^{2} 2x^3 \, dx = 2\int_{-2}^{2} x^3 \, dx = 2\left[\frac{1}{4}x^4\right]_{-2}^{2} = \frac{1}{2}\left[(2)^4 - (-2)^4\right] = 0$$

3. $\int_{1}^{2} \left(3x^2 + x + 1\right) dx$

Nothing particularly difficult about this problem, you just have to remember that it's OK to break the integral down by terms, take the antiderivatives, then solve as specified above. Rather a mess, but not hard.

$$\int_{1}^{2} \left(3x^2 + x + 1\right) dx = 3\int_{1}^{2} x^2 \, dx + \int_{1}^{2} x \, dx + \int_{1}^{2} dx$$

$$= 3\left(\frac{1}{3}x^3\right)\Big|_{1}^{2} + \frac{1}{2}x^2 \Big|_{1}^{2} + x\Big|_{1}^{2}$$

$$= \frac{3}{3}\left[(2)^3 - (1)^3\right] + \frac{1}{2}\left[(2)^2 - (1)^2\right] + [2 - 1]$$

$$= 7 + \frac{3}{2} + 1 = 9.5$$

4. Show that: $\int_{0}^{2} x^2 \, dx = \int_{0}^{1} x^2 \, dx + \int_{1}^{2} x^2 \, dx$

A simple exercise to prove to you that Rule #5 above is true. Plus you get a little more practice.

$$\int_{0}^{2} x^2 \, dx = \frac{1}{3}x^3 \Big|_{0}^{2} = \frac{1}{3}(2)^3 - \frac{1}{3}(0)^3 = \frac{8}{3}$$

$$\int_{0}^{1} x^2 \, dx = \frac{1}{3}x^3 \Big|_{0}^{1} = \frac{1}{3}(1)^3 - \frac{1}{3}(0)^3 = \frac{1}{3}$$

$$\int_{1}^{2} x^2 \, dx = \frac{1}{3}x^3 \Big|_{1}^{2} = \frac{1}{3}(2)^3 - \frac{1}{3}(1)^3 = \frac{7}{3}; \qquad \therefore \quad \frac{7}{3} + \frac{1}{3} = \frac{8}{3}$$

Additional Problems

Evaluate the given definite integrals.

1. $\int_0^3 x \, dx$

2. $\int_{-1}^2 x^3 \, dx$

3. $\int_{-5}^1 2x \, dx$

4. $\int_{-3}^{-1} 3x^2 \, dx$

5. $\int_3^{10} \left(x^2 + 5x\right) dx$

6. $\int_{-1}^3 \left(2x^3 + 3x^2 - 2x - 1\right) dx$

7. $\int_2^4 \left(x^{-2} + x^3 - x^{-3}\right) dx$

8. $\int_1^5 \left(3x^{\frac{1}{3}} + x^{\frac{1}{2}} - 2\right) dx$

9. $\int_\pi^{\frac{3\pi}{2}} \cos x \, dx$

10. $\int_{7\pi/3}^{8\pi} \cos 3x \, dx$

11. $\int_2^1 \sin x \, dx$

12. $\int_1^2 \left(\sin^2 x + \cos^2 x\right) dx$

13. $\int_3^6 \left(\frac{x^2 + 7x^2 + 2}{x^2}\right) dx$

14. $\int_{-1}^2 \left(\frac{3x^4 - 2x - 5}{x^4}\right) dx$

15. Show that: $\int_{-3}^0 x^2 \, dx = \int_{-3}^{-2} x^2 \, dx + \int_{-2}^0 x^2 \, dx$

16. Show that: $\int_{-5}^6 \left(2x^2 + 1\right) dx = \int_{-5}^0 \left(2x^2 + 1\right) dx + \int_0^6 \left(2x^2 + 1\right) dx$

5.4 Evaluating Indefinite Integrals

As we showed you in the last section, an indefinite integral is simply an integral without boundary points. The only difference in computation from definite integrals is the addition of a constant "*C*" to the expression instead of plugging in boundary points:

$$\int f(x)dx = F(x) + C$$

So why is it that we plug in this constant C here? The answer to that lies in the Fundamental Theorem of Calculus.

The Fundamental Theorem of Calculus: When you take the derivative of an integral, you get the function within the integral.

Given: $y = f(x)$, then $\dfrac{dy}{dx}\displaystyle\int f(x)dx = f(x)$

Example: $f(x) = \left(\dfrac{39x^{\frac{8}{13}}}{(x+1)\sqrt{x}}\right)$, then $\dfrac{dy}{dx}\displaystyle\int\left(\dfrac{39x^{\frac{8}{13}}}{(x+1)\sqrt{x}}\right)dx = \left(\dfrac{39x^{\frac{8}{13}}}{(x+1)\sqrt{x}}\right)$

If you think about this for a minute, it's so obvious it hurts: We take the integral of a function and get its antiderivative. We then take the derivative of that antiderivative, and get right back to where we started--the original function.

Using this reasoning, we can work back the other way--though not so neatly. We take the derivative of a function, obtaining a new function. We then take the integral of this new function, getting the original function back--*almost*. Why almost? Because the derivative wipes out constant terms. The following example illustrates this:

If we're given the function, $f(x) = 2x^3 + 5$, and we take the derivative of that function, we get , $f'(x) = 6x^2$. Now, if we take the antiderivative of $f'(x)$, we get, $f(x) = 2x^3$. What's missing? That's right, the 5 is gone. Why? Because the derivative wiped it out. Hence, we add a generic constant, C, to the answer to account for the lost constant.

Now that you understand the reasoning behind the $+ C$, we can move on to solving indefinite integrals. Fortunately for us, all of the rules we learned in regards to solving definite integrals apply here as well--with, of course, the exception mentioned above.

Examples

5.4

1. $\displaystyle\int x^2 dx$

There's really not a whole lot to say here. You integrate as before, but instead of plugging in boundary points, you just say plus C and you're done. If you got through the last section, this is trivial for you.

$\displaystyle\int x^2 dx = \dfrac{1}{3}x^3 + C$

2. $\int \left(3x^2 - 9x + 2\right)dx$

$\int \left(3x^2 - 9x + 2\right)dx = x^3 - \dfrac{9}{2}x^2 + 2x + C$

3. $\int \sin 2x \, dx$

$\int \sin 2x \, dx = -\dfrac{1}{2}\cos 2x + C$

Additional Problems   **5.4**

Evaluate the given indefinite integrals.

1. $\int x^3 dx$

2. $\int x^5 dx$

3. $\int 2x^3 dx$

4. $\int -7x^6 dx$

5. $\int \left(x^2 + 5x + 1\right)dx$

6. $\int \left(3x^5 - x^4 + 2\right)dx$

7. $\int \cos x \, dx$

8. $\int \left(\cos 3x + \sin 7x + x^2 + 1\right)dx$

9. $\int \left(x + 1\right)^3 dx$

10. $\int \left(x - 3\right)^5 dx$

11. $\int \left(\dfrac{x^2 - 2x^3 + 1}{x^2}\right)dx$

12. $\int \left(x^{\frac{1}{2}} + 3x^{\frac{1}{3}} + x^{-\frac{1}{7}}\right)dx$

5.5 The Method of Substitution

Evaluating an integral by inspection, which is what we've been doing, is sometimes easier said than done. Hence, we often have to resort to specific methods to aid us in these efforts. One such method that works in some cases is the method of substitution. For example, suppose we want to evaluate the integral:

$\int \left(2x + 3\right)^3 dx$

Solving this via the method of substitution works as follows:

Let $u = 2x + 3$

Let $du = dx$

Substituting in then, we get: $\int u^3 du$

Clearly, we can solve this integral: $\int u^3 du = \dfrac{1}{4} u^4 + C$

Substituting back to get our answer in terms of x: $\dfrac{1}{4} u^4 + C = \dfrac{1}{4}(2x + 3)^4 + C$

As you can see, essentially what we're doing here is substituting one variable for another to simplify the process of evaluating the integral. Once the integral is evaluated, we then simply substitute back to get our answer in terms of the original variable.

Examples

5.5

1. $\int (7x + 5)^8 dx$

   $u = 7x + 5$

   $du = 7\, dx \quad \therefore \quad dx = \dfrac{1}{7} du$

To solve this problem we begin by finding an obvious substitution--in this case the terms within the parentheses. Note: When looking for a substitution, make sure that upon taking the derivative of it--du, it yields a "friendly" term. Remember, the goal is to come up with an integral we can integrate!

$\int (7x + 5)^8 dx = \int u^8 \left(\dfrac{1}{7} du \right) = \dfrac{1}{7} \int u^8 du$

$= \dfrac{1}{7} \left(\dfrac{1}{9} u^9 \right) = \dfrac{1}{63}(7x + 5)^9 + C$

Once the substitution is made, the rest is easy. You just solve the easy integral, then get your answer back in terms of x by substituting back in at the end.

2. $\int \left[\dfrac{x^2 - 3}{\left(x^3 - 9x + 5 \right)^2} \right] dx$

   $u = x^3 - 9x + 5$

   $du = (3x^2 - 9)dx \quad \therefore \quad dx = \dfrac{du}{3x^2 - 9}$

In this problem, it's not so obvious what the substitution should be. Clearly, we can't account for, or cover as we say, everything in the substitution, so what we have to do is think ahead to what the derivative of u will be and see if that won't cancel out the terms we didn't cover.

For example, in this problem we can't cover the numerator, but after taking the derivative of what we did cover, we see that it will cancel out.

$$\int \left[\frac{x^2-3}{u^2}\right]\left(\frac{du}{3x^2-9}\right) = \int \left[\frac{x^2-3}{u^2}\right]\left(\frac{du}{3(x^2-3)}\right)$$

$$= \int \frac{1}{3u^2}\,du = -\frac{1}{3}u^{-1} = -\frac{1}{3}(x^3-9x+5)^{-1} + C$$

3. $\displaystyle \int \frac{x}{\sqrt[5]{x^2+5}}\,dx$

Another example showing you the need to think ahead when picking your substitution. Again, we can't cover everything in the initial substitution, but after taking the derivative of u, we find that we can cancel what we didn't cover. Pay attention to this example--this canceling technique will be used over and over.

$u = x^2 + 5$

$du = 2x\,dx \quad \therefore \quad dx = \dfrac{du}{2x}$

$$\int \frac{x}{u^{\frac{1}{5}}}\left(\frac{du}{2x}\right) = \int \frac{1}{2u^{\frac{1}{5}}}\,du = \frac{1}{2}\int u^{-\frac{1}{5}}\,du = \frac{1}{2}\left(\frac{5}{4}u^{\frac{4}{5}}\right) = \frac{5}{8}(x^2+5)^{\frac{4}{5}} + C$$

4. $\displaystyle \int \cos(5x+1)\,dx$

Another type of integral that this simple substitution works on. Note that before the substitution, solving the integral would have been most difficult. With the substitution, however, you hardly have to think.

$u = 5x + 1$

$du = 5\,dx \quad \therefore \quad dx = \dfrac{du}{5}$

$$\int \cos u\left(\frac{du}{5}\right) = \frac{1}{5}\int \cos u\,du = \frac{1}{5}(\sin u) = \frac{1}{5}\sin(5x+1) + C$$

Additional Problems   5.5
Solve the given indefinite integrals.

1. $\displaystyle \int (8x-6)^3\,dx$

2. $\displaystyle \int (17x+5)^9\,dx$

3. $\displaystyle \int \sqrt{6x+5}\,dx$

4. $\displaystyle \int \sqrt[3]{19x+96}\,dx$

5. $\displaystyle \int x\left(\sqrt{x^2+5}\right)dx$

6. $\displaystyle \int \left(\sqrt{18-3x^3}\right)(18x^2)\,dx$

7. $\displaystyle \int \sqrt{x^2-4x+4}\,dx$

8. $\displaystyle \int \frac{x}{(2x^2+5)^3}\,dx$

9. $\displaystyle \int \left[\frac{2x-3}{(x^2-3x+9)^2}\right]dx$

10. $\int \left(4x^8 - 6x^3 + 2\right)^4 \left(32x^7 - 18x^2\right)dx$

11. $\int \dfrac{35x^4 + 12x^2 - 2}{\left(7x^5 + 4x^3 - 2x + 1\right)^{\frac{1}{4}}}dx$

12. $\int \left(\dfrac{1}{\sqrt{10 + x^{-1}}}\right)\left(\dfrac{1}{x^2}\right)dx$

13. $\int \left(3\cos x - 10\right)^5 \sin x \, dx$

14. $\int 3\left(\sin^3 x\right)\left(\cos x\right)dx$

15. $\int \dfrac{\left(x^{\frac{1}{2}} + 10\right)^4}{x^{\frac{1}{2}}}dx$

16. $\int \dfrac{\cos x \sin x}{\sqrt{\cos^2 x + 10}}dx$

17. $\int \cos(10x + 13)dx$

18. $\int \sin(3x - 2)dx$

Chapter 6

Integral Applications: Areas and Volumes

6.1 Bounded Region Area Calculations
6.2 *y*-Axis Integration
6.3 Volume Calculations

In the last chapter we learned about the third and final mathematical operation that comprises calculus--the integral. In this chapter, we move on in our study of integrals to applications of integrals. Note that these applications are not real world applications like the ones we showed you in the derivative applications chapter. Rather, these are mathematical applications.

This doesn't have a good sound to it, does it? Mathematical applications? Ack! Well, actually it's not that bad. Think of this chapter as a conditioning chapter for your mind. We're in about the same situation as the average college football linemen. You see, before the season starts, most teams have their traditional mile run wherein everybody has to finish in a certain time. As you can imagine, for the linemen this is a real challenge. The 300 pounders suffer. So why do they do it? Clearly their coach makes them, right? Why does he make them? Certainly not for any practical reason. Afterall, most plays don't require the linemen to move even fifteen yards, let alone a mile. The reason they're all made to do it is for conditioning--mental if nothing else. It makes them tougher. Well, guess what? We're all about to become tougher too. Put those mental running shoes on...

6.1 Bounded Region Area Calculations

Here we go. The first of the integral applications that we're going to learn about is that of bounded region area calculations. Bounded region area calculations are used anytime you have an area surrounded by two or more functions and you need to know what it is. For example, typical bounded region area problems have two functions graphed on the same coordinate system and ask you to find the area between them for some interval (See Fig. 6.1). To solve these types of problems, we follow a quick three step procedure:

1. If the interval endpoints have not been given, determine them. Note that the only time the interval endpoints are not given is when the functions intersect and thus physically bound some area. Thus, the interval endpoints are the points where the functions intersect. To find these, simply set the functions equal to each other and solve for the *x*'s.

2. Determine which function is on top. We accomplish this by graphing the functions or by simply plugging values from the interval into the functions and comparing the results. Note that if you mess this up and solve with the wrong function on top, you'll get the same answer except it will be negative.

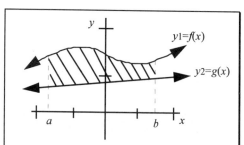

3. Plug all of this information into the following equation:

$$A = \int_a^b [\text{top function - bottom function}]$$

- -where a and b are the interval endpoints

Fig. 6.1: We have two functions graphed on the same coordinate system and are interested in the area between them on the interval $x = a$ to b.

Note: Occasionally we'll run into slightly more complicated problems with more than two functions. Not to worry. The way these problems are set up, all we have to do for these problems is divide the graph into sections such that in any one section only two functions show. We then integrate each section separately and add the results together. The examples and additional problems that follow will clarify all this.

Examples   6.1

In the following problems, find the area of the region bounded by the given functions on the given intervals.

1. $y_1 = x^2$, $y_2 = 2x + 3$, $x = 0$, $x = 2$

[1] Interval given

[2] Determine which function is on top - - clearly from the graph we see y_2 is on top. If we didn't have a graph, though, we would just plug a number from the interval into the functions to make this determination. For example:

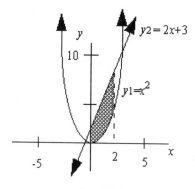

Fig. 6.2

$$f_1(1) = (1)^2 = 1$$
$$f_2(1) = 2(1) + 3 = 5$$

Since $5 > 1$, y_2 is on top.

[3] $A = \int_0^2 \left[(2x+3) - (x^2) \right] dx$

$$= \left[-\frac{1}{3}x^3 + x^2 + 3x \right]\Big|_0^2 = \left[-\frac{1}{3}(2)^3 + (2)^2 + 3(2) \right] - \left[-\frac{1}{3}(0)^3 + (0)^2 + 3(0) \right] = 7\frac{1}{3}$$

Note: If you make a mistake and put y_1 on top, look at the answer you get:

$$A = \int_0^2 \left[(x^2) - (2x+3) \right] dx = \left[\frac{1}{3}x^3 - x^2 - 3x \right]\Big|_0^2 = \left[\frac{1}{3}(2)^3 - (2)^2 - 3(2) \right] - [0] = -7\frac{1}{3}$$

2. $y_1 = 5 - x^2, \ y_2 = -2x - 3$

[1] No interval given, so we find
   the interval by setting the two
   functions equal to each other
   and solving for the x's:

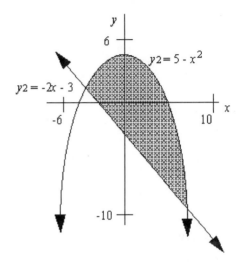

$5 - x^2 = -2x - 3$

$x^2 - 2x - 8 = 0$

$(x+2)(x-4) = 0$

$x = -2, \ x = 4$

[2] From the graph we see y_1 is on top

Fig. 6.3

[3] $A = \int_{-2}^4 \left[(5 - x^2) - (-2x - 3) \right] dx$

$$= \int_{-2}^4 \left[-x^2 + 2x + 8 \right] dx$$

$$= \left[-\frac{1}{3}x^3 + x^2 + 8x \right]\Big|_{-2}^4 = \left[-\frac{1}{3}(4)^3 + (4)^2 + 8(4) \right] - \left[-\frac{1}{3}(-2)^3 + (-2)^2 + 8(-2) \right]$$

$$= 36$$

3. $y_1 = x^2, \ y_2 = 2x^3$

[1] Find the interval: $2x^3 = x^2$

$$2x^3 - x^2 = 0$$

$$x^2(2x - 1) = 0$$

$$x = 0, \ x = \frac{1}{2}$$

[2] Decide which function is on top:

$$f_1\left(\frac{1}{4}\right) = \left(\frac{1}{4}\right)^2 = \frac{1}{16}$$

$$f_2\left(\frac{1}{4}\right) = 2\left(\frac{1}{4}\right)^3 = \frac{1}{32}$$

[3] $A = \int_0^{\frac{1}{2}}\left[x^2 - 2x^3\right]dx = \left[\frac{1}{3}x^3 - \frac{1}{2}x^4\right]\Big|_0^{\frac{1}{2}} = \left[\frac{1}{3}\left(\frac{1}{2}\right)^3 - \frac{1}{2}\left(\frac{1}{2}\right)^4 - 0\right] = 0.010$

4. Find the area of the region bounded by:

$y_1 = -2x,\ y_2 = \frac{1}{2}x,$

$y_3 = 10$

Note: We have 3 functions here. Therefore, since we only know how to solve with 2 functions at a time, divide the problem into two parts; solve for 2 areas.

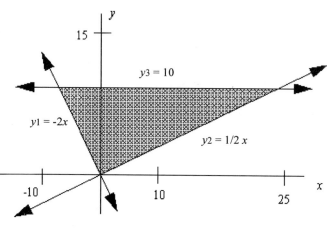

Fig. 6.4

Let A_1 equal the area of the bounded region to the left of the y_1, y_2 intersect, and let A_2 equal the area of the bounded region to the right of the y_1, y_2 intersect.

Solve for A_1

[1] Figure out the interval: $x = 0$ is one end

$$-2x = 10$$
$$x = -5 \quad \therefore \ x = -5,\ x = 0$$

[2] From the graph we see $y = 10$ is on top.

[3] $A_1 = \int_{-5}^{0}\left[10 - (-2x)\right]dx = \left[10x + x^2\right]\Big|_{-5}^{0} = \left[(0 + 0) - (-50 + 25)\right] = 25$

Solve for A_2

[1] Figure out the interval: $x = 0$ is one end

$$\frac{1}{2}x = 10$$
$$x = 20 \quad \therefore \; x = 0, \; x = 20$$

[2] From the graph we see $y = 10$ is on top.

[3] $A_2 = \int_0^{20}\left[10 - \frac{1}{2}x\right]dx = \left[10x - \frac{1}{4}x^2\right]\Big|_0^{20} = \left[\left(200 - \frac{1}{4}(20)^2\right) - 0\right] = 100$

AREA $= A_1 + A_2 = 25 + 100 = 125$

Additional Problems   **6.1**

In the following problems, find the area of the region bounded by the given functions on the given intervals.

1. $y_1 = -x^2$, $y_2 = -2x - 3$; $x = 0$, $x = 2$

4. $y_1 = -4$, $y_2 = x^2 - 10$; $x = 0$, $x = 3$

2. $y_1 = x^2 - 5$, $y_2 = 2x + 3$

5. $y_1 = -3x$, $y_2 = 5x - 5$, $y_3 = 10 - x^2$

3. $y_1 = 4$, $y_2 = 10 - x^2$; $x = 0$, $x = 3$

6.2 *y*-Axis Integration

The second mathematical application of integrals that we learn about in this chapter is y-axis integration. "What is y-axis integration?" you ask. Simply the integration of functions with respect to the y-axis instead of the x-axis--i.e., when you're integrating with respect to the y-axis, you're finding the area between the function and the y-axis. Thus, when doing y-axis integration, the bounds will be y-axis numbers, and the function will be rearranged so that it is in terms of y. This is not difficult--just mental gymnastics. Here's an example to illustrate:

Suppose we are given the function, $y = 3x + 1$, and are asked to integrate it with respect to the y-axis between $y = 2$ and 8. To solve, we follow these two simple steps:

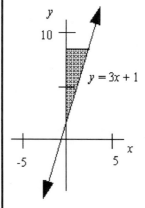

Fig. 6.5

Method for doing y-axis integration:

1. Rewrite the function in terms of y.

   Here we get: $x = \dfrac{y-1}{3}$

2. Integrate the function in its rewritten form.

   Here we get the following:

   $$\int_2^8 \left(\frac{y-1}{3}\right) dy$$

   Remember, the boundaries are on the y - axis,
   not the x - axis. See Fig. 6.5.

Often we'll be asked to find the area of the region bounded by two functions with respect to the y-axis. Here's how it's done:

1. If the interval endpoints have not been given, determine them as in the last section.

2. Determine which function is farther to the right. As in the last section, we accomplish this by either graphing or plugging values from the interval into the function.

3. Plug all of this information into the following equation:

$$A = \int_a^b [\text{function farthest to the right - the other function}] dy$$

    --where a and b are interval endpoints of the y - axis.

Examples   **6.2**

In the following problems, find the area of the function with respect to the y-axis.

1. $y = 2x + 1; \quad y = 1, \ y = 4$

[1] Get function in terms of y:

    $y = 2x + 1$

    $x = \dfrac{y-1}{2} = \dfrac{1}{2}y - \dfrac{1}{2}$

[2] $A = \int_1^4 \left[\dfrac{1}{2}y - \dfrac{1}{2}\right] dy = \left[y^2 - \dfrac{1}{2}y \right]_1^4$

    $= \left[\left((4)^2 - \dfrac{1}{2}(4) \right) - \left((1)^2 - \dfrac{1}{2}(1) \right) \right] = 13.5$

Note that there's nothing difficult about this problem. All we're doing is rearranging the function in terms of y, then integrating it. The bounds of integration were provided in the problem.

2. $y^2 + 10 = x^2 + 1 + 6y$

> *The most challenging part of this problem, as with most y-axis integration, is getting the function in terms of y. Here, things simplify greatly and the integration becomes trivial.*

[1] Get function in terms of y:

$$y^2 + 10 = x^2 + 1 + 6y$$

$$x^2 = y^2 - 6y + 9$$

$$x = \sqrt{y^2 - 6y + 9} = \sqrt{(y-3)^2} = y - 3$$

[2] $A = \int (y-3)dy = \dfrac{1}{2}y^2 - 3y + C$

3. $y = x^2 - 5$

> *The trick in this problem is to realize that, yes, you can use substitution to solve this problem. Notice once you realize this that the problem is easy.*
>
> *Remember, everything we learned last chapter about integration applies here as well.*

[1] Get function in terms of y:

$$y = x^2 - 5$$

$$x^2 = y + 5$$

$$x = \sqrt{y+5}$$

[2] $A = \int \left(\sqrt{y+5} \right) dy$

$$u = y + 5$$

$$du = dy$$

$$A = \int \left(\sqrt{u} \right) dy = \int u^{\frac{1}{2}} dy = \frac{2}{3} y^{\frac{3}{2}} = \frac{2}{3}(y+5)^{\frac{3}{2}} + C$$

4. $y = \sqrt{x+1}$

> *Here, the major stumbling block is getting the function in terms of y. Once you see how to do that, the rest is easy since the resulting integration is simple.*

[1] Get function in terms of y:

$$y = \sqrt{x+1}$$

$$y^2 = \left(\sqrt{x+1} \right)^2$$

$$y^2 = x + 1$$

$$x = y^2 - 1$$

[2] $A = \int (y^2 - 1)dy = \dfrac{1}{3}y^3 - y + C$

5. $x = 2y^2 - 4$, $x = y^2$; Find the area
the region bounded by these functions.

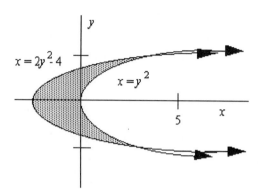

[1] Determine the boundaries:

$$2y^2 - 4 = y^2$$
$$y^2 = 4$$
$$y = -2, \ y = 2$$

[2] From the graph we can see
$x = y^2$ is farther right.

Fig. 6.6

[3] $A = \int_{-2}^{2} \left[y^2 - \left(2y^2 - 4 \right) \right] dy$

$$= \int_{-2}^{2} \left[-y^2 + 4 \right] dy = \left[-\frac{1}{3} y^3 + 4y \right]_{-2}^{2} = \left[\left(-\frac{1}{3}(2)^3 + 4(2) \right) - \left(-\frac{1}{3}(-2)^3 + 4(-2) \right) \right]$$

$$= 5.3 + 5.3 = 10.6$$

6. $x = y^2 + 1$, $x = 10$

[1] From the graph we see
$x = 10$ is farther right.

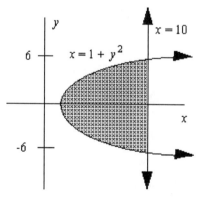

[2] Determine boundary points:

$$y^2 + 1 = 10$$
$$y^2 = 9$$
$$y = -3, \ y = 3$$

Fig. 6.7

[3] $A = \int_{-3}^{3} \left[10 - \left(y^2 + 1 \right) \right] dy$

$$= \int_{-3}^{3} \left[9 - y^2 \right] dy$$

$$= \left[9y - \frac{1}{3} y^3 \right]_{-3}^{3} = \left[\left(9(3) - \frac{1}{3}(3)^3 \right) - \left(9(-3) - \frac{1}{3}(-3)^3 \right) \right] = 18 + 18 = 36$$

Additional Problems 6.2

In the following problems, find the area of the function with respect to the *y*-axis.

1. $y = 10x - 2$

4. $y = \sqrt{x+5}$

2. $4y = 2x + 5$

5. $2y + x = 32y^2 - 18y^3 + 12$

3. $y^2 - 6 - x^2 = -10 + 4y$

6. $2y = \sqrt[3]{x} - 13$

In the following problems, find the area of the region bounded by the functions with respect to the *y*-axis.

7. $x = 5,\ x = y^2$

8. $y = \sqrt{x+5},\ y = -2x + 10$

6.3 Volume Calculations

Now the real mental gymnastics begin. It doesn't seem possible, but yes folks, we're going to take integrals, things that were made to calculate areas, and manipulate them in such a way that they calculate volumes. How? By having integrals operate on functions that represent areas rather than lines. If you notice, every integral we've worked out up to now has been operating on a function that, when graphed, represents a line. Well now we're going to take these same functions, throw them into another function that represents an area, and then integrate the whole thing. The result will be the volume produced by rotating the original function about a given axis.

Wait, it gets better. Given the ability to rotate a function about either the *x* or *y*-axis, there are four possible volumes that can be produced. See Figs. 6.8*a* and *b*.

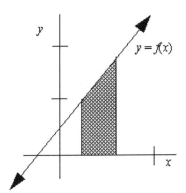

Fig. 6.8a: Shaded region can be rotated about both the *x* and *y*-axes. Thus, 2 possible volumes.

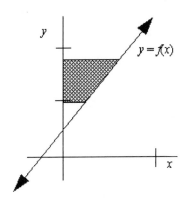

Fig. 6.8b: Shaded region can be rotated about both the *x* and *y*-axes. Thus, 2 possible volumes.

We have two methods at our disposal for calculating these volumes--the circle and cylinder methods. The circle method, sometimes referred to as the disk method, is used to find the volume produced by rotating the shaded-in region of Fig. 6.8a about the *x*-axis, and the shaded-in region of Fig. 6.8b about the *y*-axis. The cylinder method, on the other hand, is used to find the volume produced by rotating the shaded-in region of Fig. 6.8a about the *y*-axis, and the shaded-in region of Fig. 6.8b about the x-axis.

Clearly, these two methods do different jobs for us when it comes to finding the volume produced by the rotation of one function about an axis. However, when we have a region bounded by functions it doesn't matter which method we use. Both methods give us the same answer. Granted, how they come up with their answers is different, but that really doesn't matter.

So, when you're dealing with just one function, be careful. Make sure you know what is being asked so you can choose the appropriate method. When you're dealing with more than one function and the question deals with the area bounded by them, use whichever method is easier--usually you will find the circle method for *x*-axis rotation and the cylinder method for *y*-axis rotation to work best.

The Circle (Disk) Method
As we said in the introduction to this section, the way we get integrals to calculate volumes is by having them operate on functions that represent areas. Thus, the crux of both the circle and cylinder methods is the development of these functions. All you really need is the final formula for each method to start working problems, but we're going to explain how each works in detail anyway. This should prevent confusion when things get hairy. We begin with a simple example:

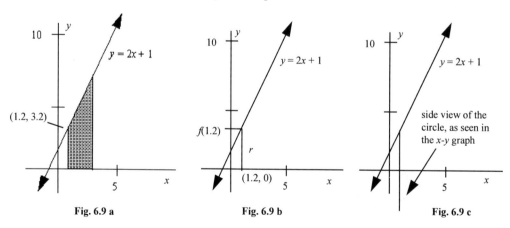

Fig. 6.9 a   Fig. 6.9 b   Fig. 6.9 c

Suppose we're given the function, $y = 2x + 1$, and told to find the volume produced upon rotation of the shaded region (Fig. 6.9 a) about the *x*-axis. Here's how we accomplish the task with the circle method:

1. We draw a vertical line from the *x*-axis to the endpoint (1.2, 3.2), Fig. 6.9b. We consider this line to be the radius of a circle (disk), origin at (1.2, 0). Note that the circle, when drawn, cannot be seen on the *x-y* graph since it is in the *y-z* plane, Fig. 6.9c.

2. We calculate the area of this circle by using the standard formula of πr^2, with the *r* being the radius found in step #1 above. *Thus, we have the area that this one point of the function makes when it is rotated about the x-axis.* We note that if we did this for every point in the function, and added the results together, we would have the total volume that the function makes upon rotation about the *x*-axis. Not wanting to waste time doing this, however, we employ an integral.

3. We set up an integral as follows:

$$V = \int_a^b \pi \left[f(x) \right]^2 dx$$

    --where *V* equals the volume, $f(x)$ the original function, and *a* and *b* the *x* - axis bounds.

    --Note that the function that this integral is operating on --$\left[f(x) \right]^2$ --is just the formula for finding the area of the circle, where $f(x)$ is analagous to *r*.

Now don't be fooled by all of this talk of *x*-axis rotation. This example focuses on *x*-axis rotation, but as we said in the introduction to this section, the circle method works for *y*-axis rotation as well. All we have to do to make the conversion is change the original function and bounds to terms of *y* as in section 6.2, and imagine the process operating off of the *y*-axis rather than the *x*-axis. Remember, though, that by doing this we'll be rotating a different area than in the case of *x*-axis rotation. Here, instead of rotating the area between the function and the *x*-axis, we'll be rotating the area between the function and the *y*-axis. DO NOT FORGET THIS!! Refer back to Figs. 6.8 *a* and *b* for a visual if you need it. The examples that follow will further clarify all this. Don't panic. Nobody gets this until they work a few problems.

One final note before going to these examples: Some problems will deal with rotating a region bounded by more than one function--such as those we found in section 6.1. To solve these problems, use the volume equation from step #3 above, with the following modifications:

For *x* - axis rotation: $\int_a^b \pi \left[\left(\text{top function} \right)^2 - \left(\text{bottom function} \right)^2 \right] dx$

For *y* - axis rotation: $\int_a^b \pi \left[\left(\text{function farthest right} \right)^2 - \left(\text{function farthest left} \right)^2 \right] dx$

Examples: Circle (Disk) Method   6.3

1. Use the Circle method to find the volume produced
    by rotating $y = 2x + 7$ from $x = -3$, $x = 5$ about
    the x - axis.

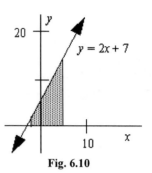

Fig. 6.10

$$V = \int_{-3}^{5} \pi [2x + 7]^2 \, dx$$

$$u = 2x + 7$$

$$du = 2 \, dx$$

$$\frac{1}{2} du = dx$$

$$V = \int_{-3}^{5} \pi [u]^2 \left(\frac{1}{2} du\right) = \frac{\pi}{2}\left(\frac{1}{3} u^3\right) = \frac{\pi}{6}(2x+7)^3 \Big|_{-3}^{5} = \frac{\pi}{6}\left[(2(5)+7)^3 - (2(-3)+7)^3\right] = 819\pi$$

2. Use the Circle method to find the volume produced
    by rotating, $y = \sqrt{x}$, from $x = 0$, $x = 4$ about the
    x - axis.

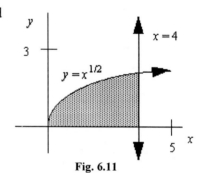

Fig. 6.11

$$V = \int_{0}^{4} \pi \left[\sqrt{x}\right]^2 \, dx$$

$$= \pi \int_{0}^{4} x \, dx = \frac{\pi}{2} x^2 \Big|_{0}^{4} = \frac{\pi}{2}\left[(4)^2 - (0)^2\right] = 8\pi$$

3. Use the Circle method to find the volume
    produced by rotating the shaded region in
    Fig. 6.12 about the y - axis.

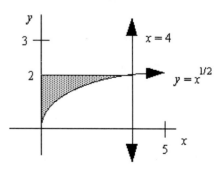

Fig. 6.12

   [1] Convert both function and interval over
        to y - terms:

$$y = \sqrt{x}$$

$$y^2 = x$$

   When $x = 0$, $y = \sqrt{0} = 0$

   When $x = 4$, $y = \sqrt{4} = 2$, -2

   --we're only interested in the $+2$ here.

$$[2] \quad V = \int_0^2 \pi \left[y^2 \right]^2 dy = \pi \int_0^2 y^4 dy = \frac{\pi}{5} y^5 \Big|_0^2 = \frac{\pi}{5} \left[(2)^5 - (0)^5 \right] = 6.4\pi$$

4. Use the Circle method to find the volume produced by rotating, $y = 2x + 7$, from $x = -3$, $x = 5$ about the y - axis.

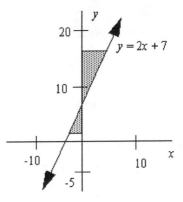

Fig. 6.13

[1] Convert everything over to terms of y:

$y = 2x + 7$

$\dfrac{y - 7}{2} = x$

   When $x = -3$, $y = 2(-3) + 7 = 1$
   When $x = 5$, $y = 2(5) + 7 = 17$

$$[2] \quad V = \int_1^{17} \pi \left[\frac{1}{2} y - \frac{7}{2} \right]^2 dy$$

$$u = \frac{1}{2} y - \frac{7}{2}$$

$$du = \frac{1}{2} dx$$

$$2 \, du = dx$$

$$V = \pi \int_1^{17} u^2 (2 du) = 2\pi \left(\frac{1}{3} u^3 \right)$$

$$= \frac{2\pi}{3} \left(\frac{1}{2} y - \frac{7}{2} \right)^3 \Big|_1^{17} = \frac{2\pi}{3} \left[\left(\frac{1}{2}(17) - \frac{7}{2} \right)^3 - \left(\frac{1}{2}(1) - \frac{7}{2} \right)^3 \right]$$

$$= \frac{2\pi}{3} \left[125 - (-27) \right] = 101.3\pi$$

5. Use the Circle method to find the volume produced by rotating the shaded region of the graph in Example 2 of this section, about the y-axis.

Note: As mentioned in the introduction to this section, this type of rotation is usually left for the Cylinder method. We can do it with the Circle method, though it is more involved.

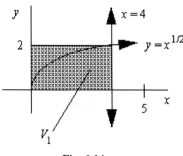

Fig. 6.14   Fig. 6.15

[1] Realize we need to do two volume calculations here:

V_1 = volume produced by rotating $x = 4$ about the y - axis (See Fig. 6.14).

V_2 = volume produced by rotating $y = \sqrt{x}$ about the y - axis (See Fig. 6.15).

[2] Convert everything over to terms of y:

$y = \sqrt{x}$

$y^2 = x$

When $x = 0$, $y = 0$

When $x = 4$, $y = +2$,   --Looking at the shaded region in Example #2, we see
we're only interested in the $+2$.

[3] $V_1 = \int_0^2 \pi (4)^2 \, dy = 16\pi \Big|_0^2 = 16\pi (2 - 0) = 32\pi$

$V_2 = \int_0^2 \pi \left(y^2\right)^2 \, dy = \pi \int_0^2 y^4 \, dy = \pi \left(\frac{1}{5} y^5\right)\Big|_0^2 = \frac{\pi}{5}\left((2)^5 - (0)^5\right) = 6.4\pi$

[4] $V = V_1 - V_2 = 32\pi - 6.4\pi = 25.6\pi$

6. Use the Circle method to find
    the volume produced by
    rotating the region bounded
    by, $y = \sqrt{x}$ and $y = \dfrac{1}{4}x$
    about the x - axis.

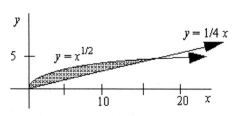

Fig. 6.16

[1] V_1 = volume produced by $y = \sqrt{x}$, rotating about the x - axis.

V_2 = volume produced by $y = \dfrac{1}{4}x$, rotating about the x - axis.

$V = V_1 - V_2$

[2] Calculate the bounds:

$$\sqrt{x} = \frac{1}{4}x$$

$$x = \frac{1}{16}x^2$$

$$x^2 - 16x = 0$$

$$x(x - 16) = 0$$

$$\therefore \ x = 0, \ x = 16$$

[3] $V = \displaystyle\int_0^{16} \pi \left[\left(\sqrt{x}\right)^2 - \left(\frac{1}{4}x\right)^2 \right] dx$

$= \displaystyle\int_0^{16} \pi \left[x - \frac{1}{16}x^2 \right] dx = \pi \left[\frac{1}{2}x^2 - \frac{1}{48}x^3 \right] \Big|_0^{16} = \pi \left[\left(\frac{1}{2}(16)^2 - \frac{1}{48}(16)^3 \right) - 0 \right] = 42.67\pi$

7. Use the Circle method to find the volume produced by rotating the region bounded by $y = \sqrt{x}$ and $y = \dfrac{1}{4}x$, about the y - axis.

[1] Convert to terms of y:

$y = \sqrt{x}$ $\qquad\qquad$ $y = \dfrac{1}{4}x$

$y^2 = x$ $\qquad\qquad$ $4y = x$

We already determined the bounds in Example #6 above for terms of x:

$x = 0, \ x = 16$

So, converting to terms of y: when $x = 0$, $y = 0$

$\qquad\qquad\qquad\qquad\qquad$ when $x = 16$, $y = 4$

[2] $V = \int_0^4 \pi\left[(4y)^2 - (y^2)^2\right]dy$, (Remember, farthest right - other)

$$= \pi \int_0^4 (16y^2 - y^4)dy = \pi\left[\frac{16}{3}y^3 - \frac{1}{5}y^5\right]\Big|_0^4 = \pi\left[\left(\frac{16}{3}(4)^3 - \frac{1}{5}(4)^5\right) - 0\right] = 136.5\pi$$

The Cylinder (Cylindrical Shells) Method

Having gotten the circle method down, we should have no problem with the cylinder method. You see, the only difference between the two is in the function that is being integrated. In the circle method the function is the area formula for a circle, while in the cylinder method the function is the area formula for a cylinder. As in the circle method, we'll introduce the formulation of this method before going on to examples. Again, we begin with a simple example:

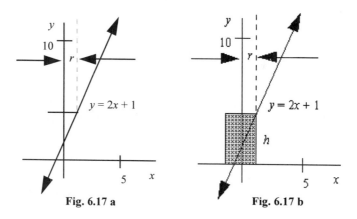

Fig. 6.17 a   Fig. 6.17 b

Suppose we're given the same function and shaded region as before in the illustration for the circle method, Fig. 6.9a, but this time we're asked to find the volume created when it's rotated about the *y*-axis. Here's how we accomplish the task with the cylinder method:

1. We rotate the endpoint of the function about the *y*-axis. This rotation produces a circular path, the distance of which can be calculated using the standard circumference of a circle: $2\pi r$. See Fig. 6.17a. Note that in Fig. 6.17a we can only see the side of the circle since the circle itself is perpendicular to the *x*-*y* plane.

2. We draw a vertical line from the *x*-axis to the point just rotated and call that the distance *h*. We then note that if we rotated this vertical line about the *y*-axis as we did the endpoint in Step #1 above, we would get a cylinder. See Fig. 6.17b.

3. To get the area of this cylinder, we simply multiply the circumference we found in Step #1 by the height just obtained in Step #2. Clearly, if we did this for every point in the function and then added the results together, we would have the total volume

produced by rotating the area between the function and the *x*-axis about the *y*-axis. Being conscious of the value of our time again, we employ an integral to do this for us.

4. We set up an integral as follows: $V = \int_a^b 2\pi x[f(x)]dx$;

--where, V is the volume, $f(x)$ is the original function, and *a* and *b* are the *x* bounds that have the following limitation: $0 \le a < b$, or, $b < a \le 0$.

--Note that the function that the integral is operating on $--2\pi x[f(x)]--$ is just the formula for the area of a cylinder.

Remember that the *x*-axis rotation is also possible with the cylinder method. To do *x*-axis rotation, simply convert everything over to terms of *y*, and imagine the method at work off of the *y*-axis. Remember, though, that in doing this a different region will be rotated. Instead of rotating the area between the function and the *x*-axis, the area between the function and the *y*-axis will be rotated. Refer back to Figs. 6.8*a* and *b* if you need a visual. The examples that follow will, as before, clarify all this.

Also, before we dive into these examples, remember that as in the circle method, some problems will require us to find the volume produced by rotating a region bounded by more than one function about a given axis. To do this, we simply use the volume equation from Step #4 above, with the following modifications:

For *x* - axis rotation: $\int_a^b 2\pi y[(\text{function farthest right}) - (\text{function farthest left})]dy$

For *y* - axis rotation: $\int_a^b 2\pi x[(\text{function on top}) - (\text{function on bottom})]dy$

Examples: Cylinder Method   **6.3**

8. Use the Cylinder method to find the volume produced by rotating $y = \sqrt{x}$ from $x = 0$, $x = 4$ about the *y* - axis. (See graph in Example 2, Fig. 6.11).

Note: We're looking for the same volume which we already solved for using the Circle method in Example #5. This will show you how much easier the Cylinder method is for this type of problem.

$$V = \int_0^4 2\pi x\left(\sqrt{x}\right)dx = 2\pi \int_0^4 x^{\frac{3}{2}}dx = 2\pi\left(\frac{2}{5}x^{\frac{5}{2}}\right)\Big|_0^4 = \frac{4\pi}{5}\left[(4)^{\frac{5}{2}} - 0\right] = 25.6\pi$$

9. Use the Cylinder method to find the volume produced
   by rotating $y = 2x + 7$, from $x = 0$, $x = 5$ about the
   y-axis (See Fig. 6.18).

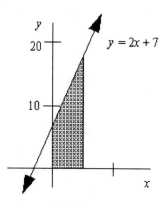

$$V = \int_0^5 2\pi x (2x + 7)dx$$

$$= 2\pi \int_0^5 (2x^2 + 7x)dx$$

$$= 2\pi \left(\frac{2}{3}x^3 + \frac{7}{2}x^2 \right) \Big|_0^5$$

$$= 2\pi \left[\left(\frac{2}{3}(5)^3 + \frac{7}{2}(5)^2 \right) - 0 \right] = 341.7\pi$$

Fig. 6.18

10. Use the Cylinder method to find the volume
    produced by rotating $y = -\frac{2}{3}x + 2$, from
    $x = -4$, $x = -1$ about the y-axis.

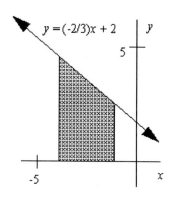

$$V = \int_{-4}^{-1} 2\pi (-x) \left[-\frac{2}{3}x + 2 \right] dx$$

$$= -2\pi \int_{-4}^{-1} \left[-\frac{2}{3}x^2 + 2x \right] dx$$

$$= -2\pi \left[-\frac{2}{9}x^3 + x^2 \right] \Big|_{-4}^{-1}$$

Fig. 6.19

$$= -2\pi \left[\left(-\frac{2}{9}(-1)^3 + (-1)^2 \right) - \left(-\frac{2}{9}(-4)^3 + (-4)^2 \right) \right]$$

$$= -2\pi \left[(1.22) - (30.22) \right] = 58\pi$$

11. Use the Cylinder method to find the volume
    produced by rotating $y = \sqrt{x}$ from $x = 0$, $x = 4$
    about the x-axis.

    [1] Convert to terms of y:

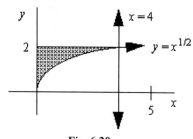

Function: Bounds:

   $y = \sqrt{x}$ When $x = 0$, $y = 0$

   $y^2 = x$ When $x = 4$, $y = +2$, -2

                         --we're only interested in $+2$

Fig. 6.20

12. Use the Cylinder method to find the volume produced by rotating $y = 2x + 7$, from $x = 0$, $x = 5$ about the x - axis.

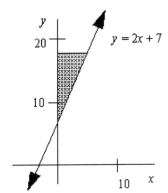

Fig. 6.21

[1] Convert to terms of y:

$$\frac{y - 7}{2} = x; \qquad \text{When } x = 0, \ y = 7$$

$$\text{When } x = 5, \ y = 2(5) + 7 = 17$$

[2] $V = \int_{7}^{17} 2\pi y \left(\frac{y - 7}{2} \right) dy$

$$= 2\pi \int_{7}^{17} \left(\frac{1}{2} y^2 - \frac{7}{2} y \right) dy$$

$$= 2\pi \left[\frac{1}{6} y^3 - \frac{7}{2} y^2 \right] \Big|_{7}^{17}$$

$$= 2\pi \left[\left(\frac{1}{6} (17)^3 - \frac{7}{2} (17)^2 \right) - \left(\frac{1}{6} (7)^3 - \frac{7}{2} (7)^2 \right) \right] = 683.2\pi$$

13. Use the Cylinder method to find the vlume produced by rotating the region bounded by $y = \sqrt{x}$ and $y = \frac{1}{4} x$, about the y - axis. See the graph in Example #6. Note: We're solving for the same volume we did in Example #7, just with a different method this time.

[1] Bounds: $\left(\frac{x}{4} \right)^2 = \left(\sqrt{x} \right)^2$

$$x^2 = 16x$$

$$x^2 - 16x = 0$$

$$x(x - 16) = 0$$

$$x = 0, \ x = 16$$

[2] $V = \int_{0}^{16} 2\pi x \left[\left(\sqrt{x} \right) - \left(\frac{1}{4} x \right) \right] dx$

$$= 2\pi \int_{0}^{16} \left(x^{\frac{3}{2}} - \frac{1}{4} x^2 \right) dx = 2\pi \left[\frac{2}{5} x^{\frac{5}{2}} - \frac{1}{12} x^3 \right] \Big|_{0}^{16} = 136.5\pi$$

14. Use the Cylinder method to find the volume produced by rotating the region bounded by $y = \dfrac{1}{4}x$ and $y = \sqrt{x}$, about the x-axis. See Fig. 6.16.

Note: We're solving for the same volume we did in Example #6.

[1] Convert to terms of y:

Function	Bounds

$y = \dfrac{1}{4}x \Rightarrow 4y = x$ When $x = 0$, $y = 0$

$y = \sqrt{x} \Rightarrow y^2 = x$ When $x = 16$, $y = 4$

[2] $V = \displaystyle\int_0^4 2\pi y\left[(4y) - (y^2)\right]dy$

$= 2\pi \displaystyle\int_0^4 \left[4y^2 - y^3\right]dy$

$= 2\pi\left[\dfrac{4}{3}y^3 - \dfrac{1}{4}y^4\right]\Big|_0^4 = 2\pi\left[\left(\dfrac{4}{3}(4)^3 - \dfrac{1}{4}(4)^4\right) - 0\right] = 42.67\pi$

Offset Axes

There will be times when we'll be asked to find the volume produced by rotating a function about some axis other than the x or y axes--i.e., an *offset axis*. This is not particularly difficult. In most cases, all we have to do is adjust the radius in each formula, and remember to subtract off excess volume to get our answer--think of offset axes problems as slightly more complex bounded region calculations.

Modifications to the radius term in the Circle Method:

1. If the offset axis is above or to the right of the function, subtract the function from the offset axis. If the offset axis is below or to the left of the function, add the function to the offset axis.

2. Use this term as the original function that you plug into the circle volume equation.

Modifications to the radius term of the Cylinder Method:

1. If the offset axis is above or to the right of the function, subtract the offset axis from the radius. If the offset axis is below or to the left of the function, add the offset axis to the radius.

2. Use this new expression as the radius that you plug into the cylinder volume equation.

In the case of bounded region volume calculations for offset axes--i.e., most problems-- the functions that bound the region must be subtracted as follows: We subtract the function closest to the offset axis from the function farthest from the offset axis.

Examples: Offset Axes 6.3

17. Use the Circle method to find the volume produced by rotating $y = x^2$ from $x = -2$, $x = 2$ about the line $y = -5$

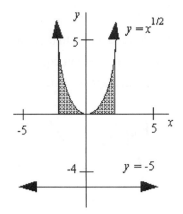

Fig. 6.23

[1] Corrected function's radius: $5 + x^2$

"Hidden" function that bounds the area: x - axis, or, $y = 5$ with respect to the offset axis.

[2] $V = \int_{-2}^{2} \pi \left[(5 + x^2)^2 - (5)^2 \right] dx$

$= \pi \int_{-2}^{2} \left[25 + 10x^2 + x^4 - 25 \right] dx$

$= \pi \left[\frac{1}{5} x^5 + \frac{10}{3} x^3 \right]_{-2}^{2} = \pi \left[\left(\frac{1}{5}(2)^5 + \frac{10}{3}(2)^3 \right) - \left(\frac{1}{5}(-2)^5 + \frac{10}{3}(-2)^3 \right) \right] = 66.2\pi$

16. Use the Circle method to find the volume produced by rotating $y = \sqrt{x}$ from $x = 0$, $x = 4$ about the line $x = 9$

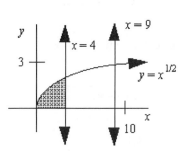

Fig. 6.24

[1] Corrected function's radius: $9 - y^2$

"Hidden" function that bounds the area: $x = 4$, but equal to $x = 5$ with respect to the offset axis.

[2] $V = \int_{0}^{2} \pi \left[(9 - y^2)^2 - (5)^2 \right] dy$

$= \int_{0}^{2} \pi \left[81 - 18y^2 + y^4 - 25 \right] dy$

$= \pi \int_{0}^{2} \left[56 - 18y^2 + y^4 \right] dy$

$= \pi \left[56y - 6y^3 + \frac{1}{5} y^5 \right]_{0}^{2} = \pi \left[\left(56(2) - 6(2)^3 + \frac{1}{5}(2)^5 \right) - 0 \right] = 70.4\pi$

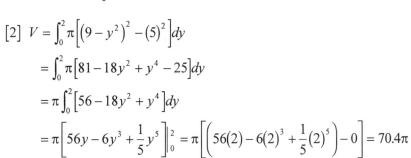

17. Use the Circle method to find the
volume produced by rotating
$y = \sqrt{x}$ from $x = 0$, $x = 4$ about
the line $x = -3$

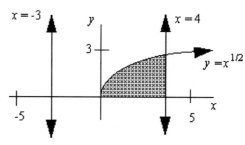

[1] Corrected function's radius: $3 + y^2$
"Hidden" function that bounds
the area: $x = -3$, but equal to
$x = 7$ with respect to the offset axis.

Fig. 6.25

[2] $V = \int_0^2 \pi \left[(7)^2 - (3 + y^2)^2 \right] dy$

$= \pi \int_0^2 \left[49 - 9 - 6y^2 - y^4 \right] dy$

$= \pi \left[40y - 2y^3 - \dfrac{1}{5} y^5 \right]\Big|_0^2 = \pi \left[\left(40(2) - 2(2)^3 - \dfrac{1}{5}(2)^5 \right) - 0 \right] = 57.6\pi$

18. Use the Cylinder method to find the
volume produced by rotating $y = \sqrt{x}$
from $x = 0$, $x = 4$ about the line $x = 5$.

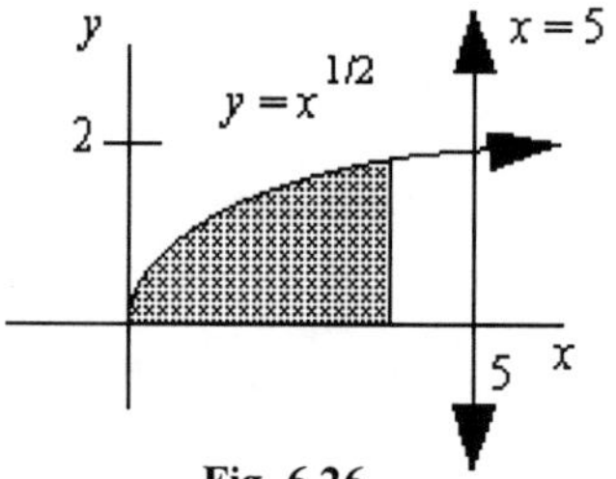

[1] Corrected radius: $5 - x$

Fig. 6.26

[2] $V = \int_0^4 2\pi (5 - x)\left[\sqrt{x} \right] dx$

$= 2\pi \int_0^4 \left[5x^{\frac{1}{2}} - x^{\frac{3}{2}} \right] dx$

$= 2\pi \left[\dfrac{10}{3} x^{\frac{3}{2}} - \dfrac{2}{5} x^{\frac{5}{2}} \right]\Big|_0^4 = 2\pi \left[\left(\dfrac{10}{3}(4)^{\frac{3}{2}} - \dfrac{2}{5}(4)^{\frac{5}{2}} \right) - 0 \right] = 27.7\pi$

19. Use the Cylinder method to find the volume produced by rotating $y = \sqrt{x}$
from $x = 0$, $x = 4$ about the line $x = -3$. See Fig. 6.25.

[1] Corrected radius: $3 + x$

[2] $V = \int_0^4 2\pi(3+x)\left[\sqrt{x}\right]dx$

$= 2\pi \int_0^4 \left[3x^{\frac{1}{2}} - x^{\frac{3}{2}}\right]dx$

$= 2\pi \left[2x^{\frac{3}{2}} - \frac{2}{5}x^{\frac{5}{2}}\right]\Big|_0^4 = 2\pi\left[\left(2(4)^{\frac{3}{2}} - \frac{2}{5}(4)^{\frac{5}{2}}\right) - 0\right] = 57.6\pi$

20. Use the Cylinder method to find the volume produced by rotating $y = \sqrt{x}$ from $x = 0$, $x = 4$ about the line $y = 8$.

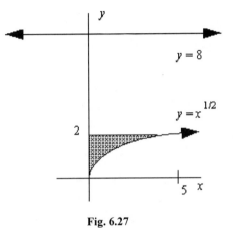

Fig. 6.27

[1] Corrected radius: $8 - y$

[2] $V = \int_0^2 2\pi(8-y)\left[y^2\right]dy$

$= 2\pi \int_0^2 \left[8y^2 - y^3\right]dy$

$= 2\pi \left[\frac{8}{3}y^3 - \frac{1}{4}y^4\right]\Big|_0^2$

$= 2\pi\left[\left(\frac{8}{3}(2)^3 - \frac{1}{4}(2)^4\right) - 0\right] = 34.67\pi$

21. Use the Cylinder method to find the volume produced by rotating $y = \sqrt{x}$ from $x = 0$, $x = 4$ about the line $y = -4$. Same shaded region as in Fig. 6.27.

[1] Corrected radius: $4 + y$

[2] $V = \int_0^2 2\pi(4+y)\left[y^2\right]dy$

$= 2\pi \int_0^2 \left[4y^2 + y^3\right]dy$

$= 2\pi \left[\frac{4}{3}y^3 - \frac{1}{4}y^4\right]\Big|_0^2 = 2\pi\left[\left(\frac{4}{3}(2)^3 - \frac{1}{4}(2)^4\right) - 0\right] = 29.33\pi$

22. Use whatever method you want to find the volume produced by rotating

the region bounded by $y = \sqrt{x}$ and $y = \dfrac{1}{4}x$ about the line $y = 10$.

We'll use the Circle method here:

[1] Corrected radius: $y = \dfrac{1}{4}x \;\Rightarrow\; 10 - \dfrac{1}{4}x$

Corrected radius: $y = \sqrt{x} \;\Rightarrow\; 10 - \sqrt{x}$

[2] $V = \displaystyle\int_0^{16} \pi \left[\left(10 - \dfrac{1}{4}x\right)^2 - \left(10 - \sqrt{x}\right)^2 \right] dx$

$= \displaystyle\int_0^{16} \pi \left[\left(100 - 5x + \dfrac{1}{16}x^2\right) - \left(100 - 20\sqrt{x} + x\right) \right] dx$

$= \displaystyle\int_0^{16} \pi \left[20\sqrt{x} - 6x + \dfrac{1}{16}x^2 \right] dx$

$= \pi \left[20\left(\dfrac{2}{3}\right)x^{\frac{3}{2}} - 3x^2 + \dfrac{1}{48}x^3 \right]\Big|_0^{16}$

$= \pi \left[\left(20\left(\dfrac{2}{3}\right)(16)^{\frac{3}{2}} - 3(16)^2 + \dfrac{1}{48}(16)^3 \right) - 0 \right] = 170.67\pi$

Additional Problems　　　　　　　　　　　　　　　　　　　6.3

1. Use the Circle method to find the volume produced by rotating $y = x^2 + 1$, from $x = 0$ to $x = 2$ about the x-axis.

2. Use the Cylinder method to find the volume produced by rotating the shaded area in Problem #1 about the y-axis.

3. Use the Circle method to find the volume produced by rotating $y = x^2 + 1$ from $x = 0$ to $x = 2$ about the y-axis.

4. Use the Cylinder method to find the volume produced by rotating the shaded region in Problem #3 about the x-axis.

5. Use the Circle method to find the volume produced by the region bounded by $y = x^2$, $y = 10 - x^2$ rotated about the y-axis.

6. Use the Cylinder method to find the volume produced by the region bounded by $y = \frac{1}{3}x^2$, $y = \sqrt{2x}$ rotated about the y - axis.

7. Use the Circle method to find the volume produced by the region bounded by $y = \frac{1}{3}x^2$, $y = \sqrt{2x}$ rotated about the y - axis.

8. Use the Circle method to find the volume produced by the region bounded by $y = \frac{1}{3}x^2$, $y = \sqrt{2x}$ rotated about the x - axis.

9. Repeat Problem #8, but this time with the Cylinder method.

10. Use the Circle method to find the volume produced by the region bounded by $y = x^2$, $y = 3 - x^2$ and the y - axis rotated about the line $y = 5$.

11. Repeat Problem #10 with the Circle method, but this time rotate the region in question about the line $y = -5$.

12. Use the Cylinder method to find the volume produced by rotating the region bounded by $y = 3x$, $y = \frac{1}{2}x^2$ about the line $x = 10$.

13. Repeat Problem #12, except this time rotate about the line $x = -2$.

14. Use the Circle method to find the volume produced by rotating the region bounded by $y = x^3$, $y = 2x$ about the line $x = 10$.

Funky Functions *plus* Special Integrals

Funky functions!?! Yes it's a pretty weird title, but then, this chapter's a bit out of the ordinary too. You see, we've hit a plateau in our study of integrals and we can't move on until we get somemore background information. Specifically, we need to learn about transcendental functions (funky functions)--i.e., non-algebraic functions like sine and cosine. Why do we need to learn about these things? Because in advanced studies of integration the integrals we're often faced with operate on transcendental functions. If we don't know anything about these types of functions, our chances of success are dismal.

Memorization. Unfortunately, that's the essence of this chapter. Except for the last two sections where we do some real integration, the majority of this chapter is spent learning about transcendental functions--which roughly translates into lots of memorization. Why all the memorization? Because each transcendental function has its own rules for what happens to it when either a derivative or an integral operates on it. For example, remember cosine and sine? How did we learn how to take their derivatives and integrals? That's right, we memorized each one's rules. Well, we have an entire chapter now of sine and cosine type functions to learn. It's not fun, but it is mandatory if we expect to succeed in the coming chapters.

7.1 Natural Logarithmic and Natural Exponential Functions

As you're all probably aware from your years in algebra, it's possible to encounter an infinite number of logarithmic and exponential functions. This being the case, you can be sure there will be many a professor asking you how to integrate them. Fortunately, as you're about to see, this is trivial. Why? Because when it comes to integrating and taking their derivatives, *all logarithmic and exponential functions fall into one of a few simple patterns. All we have to do is memorize these patterns.* In this section we'll start with the patterns for natural logarithmic and natural exponential functions, and in the next we'll finish with the ones for the general logarithmic and general exponential functions.

So, what exactly is a *natural logarithmic function?* Well, a natural logarithmic function is simply a logarithm with *base e.* It's symbol is "ln", and it's defined mathematically as:

$$\log_e x = \ln x$$

$$\ln x = \int \left(\frac{1}{x}\right) dx$$

Things to memorize about *general logarithmic functions*:

$$D_x (\ln u) = \left(\frac{1}{u}\right) D_x (u)$$

$$\int \left(\frac{1}{u}\right) du = \ln|u| + C$$

Closely related to natural logarithmic functions are *natural exponential functions.* Natural exponential functions are denoted *exp*, or more commonly just "*e*". They are defined mathematically as ln *e* = 1. Thus, natural exponential functions and natural logarithmic functions are just inverses of one another.

Things to memorize about *natural exponential functions*:

$$D_x (e^u) = e^u D_x (u)$$

$$\int e^u du = e^u + C$$

Examples: Natural Logarithms   7.1
Given $f(x)$, find $f'(x)$

1. $f(x) = \ln 2x$

$$f'(x) = \frac{1}{2x}(2) = \frac{1}{x}$$

To solve this problem all we do is follow the formula for taking the derivative of natural logarithmic functions. According to the formula, we invert the function, then multiply by the derivative of the function. Pretty simple, yes?

2. $f(x) = \ln(3x^2 + 5)$

$$f'(x) = \frac{1}{(3x^2 + 5)}(6x + 0) = \frac{6x}{(3x^2 + 5)}$$

Again, we just follow the above formula. The function is a bit more complex this time, but the idea for solving is the same. Just follow the pattern.

3. $f(x) = \ln\left(5x^3 - 2x^2 + \sqrt{3x+5}\right)$

As you can see, once you've memorized the above formula, these problems quickly become a review of derivatives. As long as you remember the rules from Chapter 3, this is a walk in the park.

$$f'(x) = \frac{1}{\left(5x^3 - 2x^2 + \sqrt{3x+5}\right)}\left(15x^2 - 4x + \frac{1}{2}(3x+5)^{-\frac{1}{2}}(3)\right)$$

$$= \frac{\left(15x^2 - 4x + \dfrac{3}{2\sqrt{3x+5}}\right)}{\left(5x^3 - 2x^2 + \sqrt{3x+5}\right)}$$

4. $f(x) = \ln\left[(7x+3)^3(8x-9)^4\right]$

The chain rule lives.

$$f'(x) = \frac{1}{(7x+3)^3(8x-9)^4}\left[3(7x+3)^2(7)(8x-9)^4 + 4(8x-9)^3(8)(7x+3)^3\right]$$

$$= \frac{21}{7x+3} + \frac{32}{8x-9}$$

5. $f(x) = \ln\left(\dfrac{x+1}{2x}\right)$

Remember the quotient rule? Here it is again: "low, D high, minus high, D low, over which the denominator squared must go".

$$f'(x) = \frac{1}{\left(\dfrac{x+1}{2x}\right)}\left[\frac{2x(1) - (x+1)(2)}{(2x)^2}\right] = \frac{2x}{x+1}\left[\frac{2x - 2x - 2}{4x^2}\right] = -\frac{1}{x(x+1)}$$

6. $f(x) = \dfrac{\sin 2x}{\ln(\sin 2x)}$

Note: We will find natural logarithmic functions buried within other functions. In this case we need to take the derivative of a fraction, and the natural log inside it. If we didn't know how to take the derivative of the natural log, we couldn't solve this problem.

$$f'(x) = \frac{[\ln(\sin 2x)]2\cos 2x - (\sin 2x)\left[\dfrac{1}{\sin 2x}(2\cos 2x)\right]}{[\ln(\sin 2x)]^2}$$

Given $f(x)$, find $\int f(x)dx$.

7. $f(x) = \dfrac{1}{2x}$

$$\int \frac{1}{2x}\,dx = \frac{1}{2}\int \frac{1}{x}\,dx = \frac{1}{2}\ln|x| + C$$

Certain functions, when integrated, have natural logs in their answers. Whenever we see a variable in the denominator raised to the first power-e.g., 1/x, think natural log. Why? Well, if you're integrating the function 1/x, ask yourself, "What function, by taking its derivative will give you 1/x?" Clearly, only ln x.

8. $f(x) = \dfrac{1}{2x+3}$

$$\int \frac{1}{2x+3}\,dx$$
$$u = 2x+3$$
$$du = 2\ dx$$
$$\int \frac{1}{u}\left(\frac{du}{2}\right) = \frac{1}{2}\ln|u| = \frac{1}{2}\ln|2x+3| + C$$

In order to solve this problem we not only need to recognize the pattern, we need to use substitution in order to get into the pattern. At first, this may seem hard. You might think, "This is impossible. I'm not going to be thinking natural log when I first try to integrate this." Actually, you will, or soon would, since this is the only way to integrate it.

9. $f(x) = \dfrac{x}{3x^2 - 5}$

$$\int \frac{x}{3x^2 - 5}\,dx$$
$$u = 3x^2 - 5$$
$$du = 6x\ dx$$
$$\int \frac{x}{u}\left(\frac{du}{6x}\right) = \frac{1}{6}\int \frac{du}{u} = \frac{1}{6}\ln|u| = \frac{1}{6}\ln|3x^2 - 5| + C$$

This problem is just like the previous one, except the function is slightly more complex, resulting in a more difficult substitution. We've seen all this before in Chapter 5, however, so this shouldn't be a big deal.

Examples: Natural Exponential Functions   7.1
Given $f(x)$, find $f'(x)$.

1. $f(x) = e^{-4x}$

$$f'(x) = e^{-4x}(-4) = -4e^{-4x}$$

Referring back to the rules for taking derivatives of natural exponential functions on page 134, we see this is simply a situation of plug 'n chug. We merely write the function down again and multiply it by the derivative of the exponent.

2. $f(x) = e^{(2x+3)}$

Again, as with natural logarithmic functions, once you start recognizing the pattern, these natural exponential problems become a review of derivatives and integrals.

$$f'(x) = e^{(2x+3)}(2+0) = 2e^{(2x+3)}$$

3. $f(x) = \left(3x + e^{4x}\right)^{\frac{1}{2}}$

A very simple problem to show you that, yes, we will find natural exponential functions buried within other functions, requiring that we know how to take their derivatives.

$$f'(x) = \frac{1}{2}\left(3x + e^{4x}\right)^{-\frac{1}{2}}\left(3 + 4e^{4x}\right)$$

4. $f(x) = e^{\ln x}$

In this problem we're merely reviewing some basic rules of logarithms and exponents. As long as you remember the rules, this is a trivial problem. See Chapter 1, pages 8 and 9, if you would like a quick review.

Since $e^{\ln x} = x$

$f'(x) = 1$

5. $f(x) = \ln e^{x}$

Again, a quick review of some basic rules of logarithms and exponents. See pages 8 an 9 in Chapter 1 for a quick review if you need one.

Since $\ln e^{x} = x$

$f'(x) = 1$

6. $f(x) = e^{\left(e^{2x}\right)}$

This type of problem is notorious in first year calculus. It's designed to trick you. Don't be a victim. Just follow through the rules of the pattern, and you'll be fine.

$$f'(x) = e^{\left(e^{2x}\right)}\left(e^{2x}\right)(2) = 2e^{2x}e^{\left(e^{2x}\right)} = 2e^{\left(2x+e^{2x}\right)}$$

7. $f(x) = 10x^{2}e^{-3x}$

Chain rule--enough said.

$$f'(x) = 20xe^{-3x} + 10x^{2}e^{-3x}(-3) = 20xe^{-3x} - 30x^{2}e^{-3x}$$

Given $f(x)$, find $\int f(x)dx$.

8. $f(x) = e^{3x}$

$\int e^{3x} dx$

$u = 3x$

$du = 3\, dx$

$\int e^u \left(\dfrac{du}{3}\right) = \dfrac{1}{3}\int e^u du = \dfrac{1}{3}e^u = \dfrac{1}{3}e^{3x} + C$

This is new, but by now you should be getting use to doing the pattern recognition thing. Here, we're just following the pattern outlined for general exponents on page 134. Same story as general logs--recognize the pattern and go. In the case of general exponents, it's easier because the e doesn't go away.

9. $f(x) = e^{(3-4x)}$

Same story as the last problem, just a slightly more complex function.

$\int e^{(3-4x)} dx$

$u = 3 - 4x$

$du = -4\, dx$

$\int e^u \left(\dfrac{du}{-4}\right) = -\dfrac{1}{4}\int e^u du = -\dfrac{1}{4}e^u = -\dfrac{1}{4}e^{(3-4x)} + C$

10. $f(x) = e^x \sin e^x$

$\int e^x \sin e^x dx$

$u = e^x$

$du = e^x dx$

$\int e^x \sin u \left(\dfrac{du}{e^x}\right) = \int \sin u \, du = -\cos u = -\cos e^x + C$

Additional Problems　　　　　　　　　　　　　　　　7.1
Given $f(x)$, find $f'(x)$.

1. $f(x) = \ln 3x$

2. $f(x) = \ln(5 - 2x)$

3. $f(x) = \ln(8x^3 + 5x^2 + 7x + 9)$

4. $f(x) = \ln\sqrt{19x + 96}$

5. $f(x) = \ln(8x^3 - 5x)^{-\frac{1}{2}}$

10. $f(x) = 4x \ln(\cos 2x)$

6. $f(x) = \ln(3x^2 + 5)^{\frac{1}{4}}$

11. $f(x) = \ln\left(\dfrac{3 - x^2}{9x + 1}\right)$

7. $f(x) = \ln\left[(13x + 1)(8x^2 - 9x)^3\right]$

12. $f(x) = 17x^3 - 1 + \ln(5x)$

8. $f(x) = \ln\left[7x^2 - (x + 5)^{\frac{1}{3}}\right]$

13. $f(x) = \dfrac{\sin 4x}{\ln(3x^4)}$

9. $f(x) = \ln(\sin 3x)$

14. $f(x) = x^2(\sin 3x)[\ln(\sin 3x)]$

Given $f(x)$, find $\int f(x)dx$.

15. $f(x) = \dfrac{1}{8x}$

18. $f(x) = \dfrac{x - 3}{x^2 - 9}$

16. $f(x) = \dfrac{1}{7x + 5}$

19. $f(x) = \dfrac{3 + \ln x}{x}$

21. $f(x) = \dfrac{\sin x}{\cos x}$

17. $f(x) = \dfrac{x}{8x^2 + 5}$

20. $f(x) = \dfrac{\cos x}{\sin x}$

Given $f(x)$, find $f'(x)$.

22. $f(x) = e^{2x}$

25. $f(x) = e^{-(19x + 96)}$

28. $f(x) = \ln e^{3x^2}$

23. $f(x) = e^{-8x^3}$

26. $f(x) = 18e^{(x^2 - 1)} + 2x^3 + 1$

29. $f(x) = \cos e^{2x}$

24. $f(x) = e^{(19x + 96)}$

27. $f(x) = (8x + e^{2x})^{\frac{1}{3}}$

30. $f(x) = x^2 \sin e^{3x}$

Given $f(x)$, find $\int f(x)dx$.

31. $f(x) = e^{2x}$

33. $f(x) = e^{(8x-7)}$

35. $f(x) = \dfrac{e^x}{\left(e^x + 2\right)^2}$

32. $f(x) = x^2 e^{-8x^3}$

34. $f(x) = 10e^{2x}\cos e^{2x}$

7.2 General Logarithmic and General Exponential Functions

Of the patterns that we're showing you for logarithmic and exponential functions in these two sections, the ones we're about to show you are by far the more powerful. Why? Because the patterns for the natural logarithms and natural exponentials can be derived from the ones in this section. If you think about it, this makes sense since this section is for the general case. Memorize these patterns.

We'll start with the general logarithmic functions. What exactly do we mean by general logarithmic functions? We mean a logarithm to any positive real number base. Mathematically we can represent this as:

$y = \log_a u$

--where $a > 0$

Things to memorize about *general logarithmic functions*:

$$D_x \log_a u = \left(\frac{1}{u \ln a}\right) D_x u$$

Closely related to the general logarithm functions are the general exponential functions. General exponential functions are simply represented as "*a*", where "*a*" is any positive real number.

Things to memorize about *general exponential functions*:

$$a^u = e^{u \ln a} = e^{\ln a^u}$$

$$D_x a^u = \left(a^u \ln a\right) D_x u$$

$$\int a^u du = \left(\frac{1}{\ln u}\right) a^u + C$$

Now don't get psyched out by the complexity of these formulas. Yes, it's true they're not that nice to look at, but once you get into some problems you'll find them to be easy to work with. Observe...

Examples 7.2

Given $f(x)$, find $f'(x)$.

1. $f(x) = \log_2(x^2 + 3)$

Follow the formula on page 141 to solve: Invert the stuff inside the parentheses times the natural log of the base, then multiply the whole thing by the derivative of the stuff inside the log.

$$f'(x) = \left(\frac{1}{(x^2+3)\ln 2}\right)(2x) = \frac{2x}{(x^2+3)\ln 2}$$

2. $f(x) = \log_3(3x^2 + 5)^3$

Same type of problem as the one above, just a slightly more complex function.

$$f'(x) = \left(\frac{1}{(3x^2+5)^3 \ln 3}\right)\left[3(3x^2+5)^2(6x)\right] = \frac{18x}{(3x^2+5)\ln 3}$$

3. $f(x) = \log_2 \ln x$

Don't be fooled here by the mixing of the logs. Just follow along with the rules for the general log, plug into the formula, and chug.

$$f'(x) = \frac{1}{(\ln x)(\ln 2)}\left(\frac{1}{x}\right) = \frac{1}{x(\ln x)(\ln 2)}$$

4. $f(x) = 3x^4 \log_5(x^2 + 5x)$

As with the natural logs, there will be problems involving general logs embedded in the functions being operated on.

$$f'(x) = 12x^3 \log_5(x^2 + 5x) + 3x^4\left[\frac{1}{(x^2+5x)\ln 5}(2x+5)\right]$$

5. $f(x) = 8^x$

Our first general exponential problem. To solve, we simply look back to the formulas on page 140. Here, we have a very simple function, so there's not much to it. We rewrite the function, multiply in by the natural log of the base, then multiply all that by the derivative of the exponent.

$$f'(x) = (8^x \ln 8)(1) = 8^x \ln 8$$

6. $f(x) = 19^{(x^2+5)}$
   Same as the previous problem, you simply look at the formula and follow the steps.

$$f'(x) = 19^{(x^2+5)}(\ln 19)(2x)$$

7. $f(x) = (x+2)^{x^2}$
   General exponents are sometimes a bit disguised, as this one is. Don't be fooled.

$$f'(x) = (x+2)^{x^2}\left(\ln(x+2)\right)(2x)$$

Given $f(x)$, find $\int f(x)dx$.

8. $f(x) = 8^x$
   Our first general exponential function integration problem. This example is just an exercise in translation--if you can read the formula, you can solve the problem.

$$\int 8^x dx = \left(\frac{1}{\ln 8}\right)8^x + C$$

9. $f(x) = x^3 3x^4$

$$\int x^3 3^{x^4} dx$$

$$u = x^4$$

$$du = 4x^3 dx$$

A much more complex problem, in order to solve this we have to recognize the pattern and use substitution to get into the correct form.

$$\int x^3 3^u\left(\frac{du}{4x^3}\right) = \frac{1}{4}\int 3^u du = \frac{1}{4}\left(\frac{1}{\ln 3}\right)3^u = \frac{1}{4}\left(\frac{1}{\ln 3}\right)3^{x^4} + C$$

10. $f(x) = 2^x(\ln 2)(\sin 2^x)$

$$\int 2^x(\ln 2)(\sin 2^x)dx$$

$$u = 2^x$$

$$du = 2^x(\ln 2)(1)dx$$

$$\frac{du}{2^x \ln 2} = dx$$

$$\int 2^x(\ln 2)(\sin u)\left(\frac{du}{2^x \ln 2}\right) = \int \sin u\, du = -\cos u = -\cos 2^x + C$$

Of special note in this problem is the selective substitution we did. Note that we only substituted u in at one of the two possible locations. This is legal. Use this trick whenever you find it convenient.

Additional Problems   **7.2**

Given $f(x)$, find $f'(x)$.

1. $f(x) = \log_3(x^3 + 4)$

7. $f(x) = 19^{(x-1)^2}$

2. $f(x) = \log_{12}(x^7 - x^3 + \sqrt{23x})$

8. $f(x) = (x^2 + 1)^{(x-1)}$

3. $f(x) = \log_2 \sin(8x^3 + 5)$

9. $f(x) = (19x + 96)^{3x^2}$

4. $f(x) = \log_4(2x - 3)^{\frac{1}{3}}$

10. $f(x) = 3^{\sin x}$

5. $f(x) = 8x^2 \log_2(3x^3 - 5x^2 + 1)$

11. $f(x) = 8^{(\cos^2 x)}$

6. $f(x) = 3^x$

12. $f(x) = (\sin 3x)^{2x}$

Given $f(x)$, find $\int f(x)dx$.

13. $f(x) = 3^x$

16. $f(x) = x^2 2^{x^3}$

14. $f(x) = 8^{5x}$

17. $f(x) = 4^x(\ln 4)\cos 4^x$

15. $f(x) = 8^{-3x}$

18. $f(x) = \dfrac{1}{2x \log x}$

7.3 Trigonometric Functions

Now that our memorization skills are warmed up, it's time to move on to more serious work. The next three sections are basically just lists. MEMORIZE THEM. Their contents will prove invaluable in the near future--Chapter 8. We begin this section with two lists concerning trigonometric functions.

Things to memorize about *Trigonometric Functions*:

Derivatives of Trigonometric Functions

1. $D_x \sin u = (\cos u)D_x u$

2. $D_x \cos u = (-\sin u)D_x u$

3. $D_x \tan u = (\sec^2 u)D_x u$

4. $D_x \csc u = (-\csc u)(\cot u)D_x u$

5. $D_x \sec u = (\sec u)(\tan u)D_x u$

6. $D_x \cot u = (-\csc^2 u)D_x u$

Integrals of Trigonometric Functions

1. $\int \sin u \, du = -\cos u + C$

2. $\int \cos u \, du = \sin u + C$

3. $\int \tan u \, du = \ln|\sec u| + C$

4. $\int \cot u \, du = \ln|\sin u| + C$

5. $\int \sec u \, du = \ln|\sec u + \tan u| + C$

6. $\int \csc u \, du = \ln|\csc u - \cot u| + C$

7. $\int \sec^2 u \, du = \tan u + C$

8. $\int \csc^2 u \, du = -\cot u + C$

9. $\int (\sec u)(\tan u) \, du = \sec u + C$

10. $\int (\csc u)(\cot u) \, du = -\csc u + C$

Examples   **7.3**

Given $f(x)$, find $f'(x)$.

1. $f(x) = \tan(14x^2)$

In all of these problems, the name of the game is just following the above formulas to solve the problems. No explanation required.

$$f'(x) = \left[\sec^2(14x^2)\right]28x = 28x\sec^2(14x^2)$$

2. $f(x) = \csc(10x^3 - 2x + 1)$

$$f'(x) = \left[-\csc(10x^3 - 2x + 1)\right]\left[\cot(10x^3 - 2x + 1)\right]\left[30x^2 - 2\right]$$

3. $f(x) = x^2 \cot(3x^3)$ *The chain rule...*

$$f'(x) = 2x\cot(3x^3) + x^2\left[-\csc^2(3x^3)\right]\left[9x^2\right]$$

4. $f(x) = \ln\left(\sec\left(8x^4\right)\right)$

$$f'(x) = \frac{1}{\sec\left(8x^4\right)}\left[\sec\left(8x^4\right)\right]\left[\tan\left(8x^4\right)\right]\left[32x^3\right]$$

5. $f(x) = \tan^3\left(8x^2\right)$

$$f'(x) = 3\left[\tan\left(8x^2\right)\right]^2\left[\sec^2\left(8x^2\right)\right]\left[16x\right] = 48x\left[\tan\left(8x^2\right)\right]^2\left[\sec^2\left(8x^2\right)\right]$$

Given $f(x)$, find $\int f(x)dx$.

6. $f(x) = x\tan\left(14x^2\right)$

$$\int x\tan\left(14x^2\right)dx$$

$$u = 14x^2$$
$$du = 28x\ dx$$

$$\int x\tan u\left(\frac{du}{28x}\right) = \frac{1}{28}\int \tan u\ du = \frac{1}{28}\ln|\sec u| = \frac{1}{28}\ln\left|\sec\left(14x^2\right)\right| + C$$

7. $f(x) = \tan 4x + \cot 9x$

$$\int\left(\tan 4x + \cot 9x\right)dx = \int \tan 4x\ dx + \int \cot 9x\ dx$$

$$u = 4x \qquad \text{and} \qquad u = 9x$$
$$du = 4\ dx \qquad\qquad du = 9\ dx$$

$$\int \tan u\ \frac{du}{4} + \int \cot u\ \frac{du}{9}$$

$$= \frac{1}{4}\ln|\sec u| + \frac{1}{9}\ln|\sin u| = \frac{1}{4}\ln|\sec 4x| + \frac{1}{9}\ln|\sin 9x| + C$$

8. $f(x) = \dfrac{1 + \sin x}{\cos^2 x}$

$$\int \left(\frac{1 + \sin x}{\cos^2 x} \right) dx = \int \frac{1}{\cos^2 x} dx + \int \frac{\sin x}{\cos^2 x} dx$$

$$= \int \sec^2 x \, dx + \int \tan x \sec x \, dx = \tan x + \sec x + C$$

9. $f(x) = \dfrac{\tan^2 8x}{\sec 8x}$

$$\int \left(\frac{\tan^2 8x}{\sec 8x} \right) dx = \int \frac{\left(\dfrac{\sin^2 8x}{\cos^2 8x} \right)}{\left(\dfrac{1}{\cos 8x} \right)} dx$$

$$= \int \left(\frac{\sin^2 8x}{\cos 8x} \right) dx$$

$$= \int \frac{1 - \cos^2 8x}{\cos 8x} dx$$

$$= \int \frac{1}{\cos 8x} dx + \int \cos 8x \, dx$$

$$u = 8x \qquad \text{and} \qquad u = 8x$$
$$du = 8dx \qquad\qquad\qquad du = 8dx$$

$$\int \frac{1}{\cos u} \left(\frac{du}{8} \right) + \int \cos u \left(\frac{du}{8} \right) = \int \sec u \left(\frac{du}{8} \right) + \int \cos u \left(\frac{du}{8} \right) = \ln|\sec u + \tan u| - \frac{1}{8} \sin u$$

$$= \ln|\sec 8x + \tan 8x| - \frac{1}{8} \sin 8x + C$$

Additional Problems   7.3
Given $f(x)$, find $f'(x)$.

1. $f(x) = \tan 2x$

2. $f(x) = \csc(-8x^3)$

3. $f(x) = \sec\left[(x^2 - 1)^2 \right]$

4. $f(x) = \cot 8x$

5. $f(x) = \csc(9x^2 + 2x - 10)$

6. $f(x) = 8x^4 \tan(3x^4)$

7. $f(x) = \cot^5(9x + 1)$

Given $f(x)$, find $\int f(x)dx$.

8. $f(x) = \tan 1996x$

9. $f(x) = \cot(8x+1)$

10. $f(x) = x^2 \sec x^3$

11. $f(x) = \dfrac{10}{\sin x}$

12. $f(x) = (\csc^2 2x)(\tan^2 2x)$

13. $f(x) = \dfrac{1+\tan^3 x}{\sin^2 x}$

14. $f(x) = 34(\tan x)(\sec x)$

15. $f(x) = \dfrac{\cos x + 1}{\sin^2 x}$

7.4 Inverse Trigonometric Functions

From your years in geometry and trigonometry, you're undoubtedly familiar with the inverse trigonometric functions. Here we provide two lists concerning them that you need to know.

Things to memorize *about Inverse Trigonometric Functions*:

Derivatives of Inverse Trigonometric Functions

1. $D_x \sin^{-1} u = \dfrac{1}{\sqrt{(1-u^2)}} D_x u$

2. $D_x \cos^{-1} u = -\dfrac{1}{\sqrt{(1-u^2)}} D_x u$

3. $D_x \tan^{-1} u = \dfrac{1}{(1+u^2)} D_x u$

4. $D_x \sec^{-1} u = \dfrac{1}{u\sqrt{(1+u^2)}} D_x u$

Integrals of Inverse Trigonometric Functions

1. $\int \dfrac{1}{\sqrt{1-u^2}} du = \sin^{-1} u + C$

2. $\int \dfrac{1}{(1+u^2)} du = \tan^{-1} u + C$

3. $\int \dfrac{1}{u\sqrt{1-u^2}} du = \sec^{-1} u + C$

4. $\int \dfrac{1}{\sqrt{1+u^2}} du = \sin^{-1}\left(\dfrac{u}{a}\right) + C$

5. $\int \dfrac{1}{(a^2+u^2)} du = \dfrac{1}{a}\tan^{-1}\left(\dfrac{u}{a}\right) + C$

6. $\int \dfrac{1}{u\sqrt{u^2-a^2}} du = \dfrac{1}{a}\sec^{-1}\left(\dfrac{u}{a}\right) + C$

Examples 7.4

Given $f(x)$, find $f'(x)$.

1. $f(x) = \sin^{-1}(8x)$

As in the previous section, we're just following the formula here, and plugging in.

$$f'(x) = \frac{1}{\sqrt{1-(8x)^2}}(8) = \frac{8}{\sqrt{1-64x^2}}$$

2. $f(x) = \cos^{-1}(3x^3 + x^{-2})$

$$f'(x) = \frac{-1}{\sqrt{1-(3x^3+x^{-2})^2}}\frac{1}{2}(9x^2 - 2x^{-3}) = -\frac{(9x^2 - 2x^{-3})}{\sqrt{1-(3x^3+x^{-2})^{\frac{1}{2}}}}$$

3. $f(x) = 2x^2 \tan^{-1}\left(x^{\frac{1}{3}}\right)$

$$f'(x) = 4x \tan^{-1}\left(x^{\frac{1}{3}}\right) + 2x^2 \left[\frac{1}{1+\left(x^{\frac{1}{3}}\right)}\right]\left(\frac{1}{3}x^{-\frac{2}{3}}\right)$$

$$= 4x \tan^{-1}\left(x^{\frac{1}{3}}\right) + \frac{2}{3}x^{\frac{4}{3}}\left(\frac{1}{1+x^{\frac{2}{3}}}\right)$$

4. $f(x) = \left(\sec^{-1}4x\right)^3$

$$f'(x) = 3\left(\sec^{-1}4x\right)^2\left[\frac{1}{4x\sqrt{(4x)^2-1}}\right](4) = 12\left(\frac{\left(\sec^{-1}4x\right)^2}{4x\sqrt{(4x)^2-1}}\right)$$

Given $f(x)$, find $\int f(x)dx$.

5. $f(x) = \dfrac{1}{25 + x^2}$

$$\int \dfrac{1}{25 + x^2}dx = \int \dfrac{1}{(5)^2 + x^2}dx = \dfrac{1}{5}\tan^{-1}\left(\dfrac{x}{5}\right) + C$$

6. $f(x) = \dfrac{x}{\sqrt{1 - x^4}}$

$$\int \dfrac{x}{\sqrt{1 - x^4}}\,dx = \int \dfrac{x}{\sqrt{1 - (x^2)^2}}\,dx$$

$$u = x^2$$

$$du = 2x\,dx$$

$$\int \dfrac{x}{\sqrt{1 - u^2}}\left(\dfrac{du}{2x}\right) = \dfrac{1}{2}\int \dfrac{1}{\sqrt{1 - u^2}}\,du = \dfrac{1}{2}\sin^{-1}u = \dfrac{1}{2}\sin^{-1}(x^2) + C$$

7. $f(x) = \dfrac{1}{x\sqrt{x^8 - 9}}$

$$\int \dfrac{1}{x\sqrt{x^8 - 9}}\,dx = \int \dfrac{1}{x\sqrt{(x^4)^2 - (3)^2}}\,dx$$

$$u = x^4$$

$$du = 4x^3\,dx$$

$$\int \dfrac{1}{x\sqrt{u^2 - (3)^2}}\left(\dfrac{du}{4x^3}\right) = \dfrac{1}{4}\int \dfrac{1}{x^4\sqrt{u^2 - (3)^2}}\,du$$

$$= \dfrac{1}{4}\int \dfrac{1}{u\sqrt{u^2 - (3)^2}}\,du = \dfrac{1}{4}\left[\dfrac{1}{3}\sec^{-1}\left(\dfrac{u}{3}\right)\right] = \dfrac{1}{12}\sec^{-1}\left(\dfrac{x^4}{3}\right) + C$$

8. $f(x) = \dfrac{1}{(x+2)\sqrt{x}}$

$$\int \frac{1}{(x+2)\sqrt{x}}\,dx = \int \frac{1}{\left[\left(\sqrt{x}\right)^2 + \left(\sqrt{2}\right)^2\right]\sqrt{x}}\,dx$$

$$u = \sqrt{x} = x^{\frac{1}{2}}$$

$$du = \frac{1}{2}x^{-\frac{1}{2}}\,dx$$

$$\int \frac{1}{\left[u^2 + \left(\sqrt{2}\right)^2\right]\sqrt{x}}\left(\frac{du}{\frac{1}{2}x^{-\frac{1}{2}}}\right) = \int \frac{1}{\left[u^2 + \left(\sqrt{2}\right)^2\right]\sqrt{x}}\left(2\sqrt{x}\right)du$$

$$= 2\int \frac{1}{\left[u^2 + \left(\sqrt{2}\right)^2\right]}\,du = \frac{2}{\sqrt{2}}\tan^{-1}\left(\frac{u}{\sqrt{2}}\right) = \frac{2}{\sqrt{2}}\tan^{-1}\left(\frac{\sqrt{x}}{\sqrt{2}}\right) + C$$

Additional Problems   7.4

Given $f(x)$, find $f'(x)$.

1. $f(x) = \sin^{-1}3x$

2. $f(x) = \cos^{-1}\left(18x^2 + 3x^{-2}\right)$

3. $f(x) = \tan^{-1}\left(\sqrt{x+1}\right)$

4. $f(x) = \sec^{-1}11x$

5. $f(x) = 4x^3\cos^{-1}\left(81x^2 - 3\right)$

Given $f(x)$, find $\int f(x)\,dx$.

6. $f(x) = \dfrac{1}{x^2 + 9}$

7. $f(x) = \dfrac{1}{(x+3)\sqrt{x}}$

8. $f(x) = \dfrac{1}{\sqrt{1-16x^2}}$

9. $f(x) = \dfrac{1}{\sqrt{e^{2x} - 81}}$

10. $f(x) = \dfrac{\sin x}{\sqrt{16 - \cos^2 x}}$

7.5 Hyperbolic Functions

Unlike the functions from the last two sections, you've probably never heard of hyperbolic functions before. Since they usually aren't seen outside of certain engineering and applied science applications, this is not surprising. Fortunately, it is not our purpose to delve into the origin, purpose, theory, etc., of hyperbolic functions. We merely need to know their mathematical makeup and how to find their derivatives, integrals, and inverses. So without further adieu, let's begin the last of the memorization processes of this chapter.

Things you need to memorize about *Hyperbolic Functions*:

Definitions of Hyperbolic Functions

1. $\sinh x = \dfrac{e^x - e^{-x}}{2}$

4. $\coth x = \dfrac{\cosh x}{\sinh x} = \dfrac{e^x + e^{-x}}{e^x - e^{-x}}$

2. $\cosh x = \dfrac{e^x + e^{-x}}{2}$

5. $\operatorname{sech} x = \dfrac{1}{\cosh x} = \dfrac{1}{e^x + e^{-x}}$

3. $\tanh x = \dfrac{\sinh x}{\cosh x} = \dfrac{e^x - e^{-x}}{e^x + e^{-x}}$

4. $\operatorname{csch} x = \dfrac{1}{\sinh x} = \dfrac{1}{e^x - e^{-x}}$

Rules: $\cosh^2 x + \sinh^2 x = 1$; $\operatorname{sech}^2 x = 1 - \tanh^2 x$; $\operatorname{csch}^2 x = \coth^2 x - 1$

Derivatives of Hyperbolic Functions

1. $D_x \sinh u = (\cosh u) D_x u$

2. $D_x \cosh u = (\sinh u) D_x u$

3. $D_x \tanh u = (\operatorname{sech}^2 u) D_x u$

4. $D_x \coth u = (-\operatorname{csch}^2 u) D_x u$

5. $D_x \operatorname{sech} u = (-\operatorname{sech} u)(\tanh u) D_x u$

6. $D_x \operatorname{csch} u = (-\operatorname{csch} u)(\coth u) D_x u$

Integrals of Hyperbolic Functions

1. $\displaystyle \int \sinh u \, du = \cosh u + C$

2. $\displaystyle \int \cosh u \, du = \sinh u + C$

3. $\displaystyle \int \operatorname{sech}^2 u \, du = \tanh u + C$

4. $\displaystyle \int \operatorname{csch}^2 u \, du = -\coth u + C$

5. $\displaystyle \int (\operatorname{sech} u)(\tanh u) \, du = -\operatorname{sech} u + C$

6. $\displaystyle \int (\operatorname{csch} u)(\coth u) \, du = -\operatorname{csch} u + C$

Examples 7.5

Given $f(x)$, find $f'(x)$.

1. $f(x) = \sinh(2x)$

$f'(x) = [\cosh(2x)](2)$

2. $f(x) = \cosh(x^2)$

$f'(x) = [\sinh(x^2)](2x)$

3. $f(x) = \tanh\left[(x + x^3)^{\frac{1}{3}}\right]$

$f'(x) = \left[\sec h(x + x^3)^{\frac{1}{3}}\right]^2 \left[\frac{1}{3}(x + x^3)^{-\frac{2}{3}}(1 + 3x^2)\right]$

4. $f(x) = \coth(x^2 + 1)$

$f'(x) = \left[-\csc h^2(x^2 + 1)\right](2x)$

5. $f(x) = 3x^4 \sec h(2x + x^3 - x^{-2})$

$f'(x) = 12x^3 \sec h(2x + x^3 - x^{-2})$
$\qquad + 3x^4\left[-\sec h(2x + x^3 - x^{-2})\right]\left[\tanh(2x + x^3 - x^{-2})\right]\left[2 + 3x^2 + 2x^{-3}\right]$

Given $f(x)$, find $\int f(x)dx$.

6. $f(x) = \sinh(2x)$

$\int \sinh(2x)dx$
$\qquad u = 2x$
$\qquad du = 2\ dx$
$\int \sinh u\left(\frac{du}{2}\right) = \frac{1}{2}\cosh u = \frac{1}{2}\cos(2x) + C$

7. $f(x) = \sinh x \cosh x$

$\int \sinh x \cosh x \, dx$

$\quad u = \sinh x$

$\quad du = \cosh x \, dx$

$\int u \cosh x \left(\dfrac{du}{\cosh x} \right) = \int u \, du = \dfrac{1}{2} u^2 = \dfrac{1}{2} (\sinh x)^2 + C$

8. $f(x) = \dfrac{\sinh x}{\cosh^2 x}$

$\int \dfrac{\sinh x}{\cosh^2 x} \, dx = \int \left(\dfrac{\sinh x}{\cosh x} \right) \left(\dfrac{1}{\cosh x} \right) dx = \int (\tanh x)(\sec hx) dx = -\sec hx + C$

Things to memorize about *Inverse Hyperbolic Functions*:

Definitions of Inverse Hyperbolic Functions

1. $\sinh^{-1} x = \ln\left[x + \sqrt{x^2 + 1} \right]$

3. $\tanh^{-1} x = \dfrac{1}{2}\left(\ln\left[\dfrac{1+x}{1-x} \right] \right); \ |x| < 1$

2. $\cosh^{-1} x = \ln\left[x + \sqrt{x^2 - 1} \right]$

4. $\sec h^{-1} x = \ln\left[\dfrac{1 + \sqrt{1 - x^2}}{x} \right]; \ 0 < x \le 1$

Derivatives of
Inverse Hyperbolic Functions

1. $D_x \sinh^{-1} u = \dfrac{1}{\sqrt{u^2 + 1}} D_x u$

2. $D_x \cosh^{-1} u = \dfrac{1}{\sqrt{u^2 - 1}} D_x u$

3. $D_x \tanh^{-1} u = \dfrac{1}{1 - u^2} D_x u$

4. $D_x \sec h^{-1} u = \dfrac{1}{u\sqrt{1 - u^2}} D_x u$

Integrals of
Inverse Hyperbolic Functions

1. $\displaystyle\int \dfrac{1}{\sqrt{u^2 + a^2}} \, du = \sinh^{-1}\left(\dfrac{u}{a} \right) + C$

2. $\displaystyle\int \dfrac{1}{\sqrt{u^2 - a^2}} \, du = \cosh^{-1}\left(\dfrac{u}{a} \right) + C$

3. $\displaystyle\int \dfrac{1}{a^2 - u^2} \, du = \dfrac{1}{a}\tanh^{-1}\left(\dfrac{u}{a} \right) + C$

4. $\displaystyle\int \dfrac{1}{u\sqrt{a^2 - u^2}} \, du = -\dfrac{1}{a}\sec h^{-1}\left(\dfrac{|u|}{a} \right) + C$

Examples

Given $f(x)$, find $f'(x)$.

1. $f(x) = \sinh^{-1}(2x)$

$$f'(x) = \frac{1}{\sqrt{(2x)^2 + 1}}(2) = \frac{2}{\sqrt{(2x)^2 + 1}}$$

2. $f(x) = \cosh^{-1}(x^3)$

$$f'(x) = \frac{1}{\sqrt{(x^3)^2 - 1}}(3x^2) = \frac{3x^2}{\sqrt{x^6 - 1}}$$

3. $f(x) = \sec h^{-1}(x + 1)$

$$f'(x) = -\frac{1}{(x+1)\sqrt{1-(x+1)^2}}(1) = -\frac{1}{(x+1)\sqrt{1-(x+1)^2}}$$

Given $f(x)$, find $\int f(x)dx$.

4. $f(x) = \dfrac{1}{\sqrt{(x+3)(x-3)}}$

$$\int \frac{1}{\sqrt{(x+3)(x-3)}}\, dx = \int \frac{1}{\sqrt{x^2 - 9}}\, dx = \int \frac{1}{\sqrt{x^2 - (3)^2}}\, dx = \cosh^{-1}\left(\frac{x}{3}\right) + C$$

5. $f(x) = \dfrac{x^{-\frac{1}{2}}}{\sqrt{2 + x}}$

$$\int \frac{x^{-\frac{1}{2}}}{\sqrt{2+x}}\, dx = \int \frac{1}{(\sqrt{x})\sqrt{(\sqrt{2})^2 + (\sqrt{x})^2}}\, dx$$

$$u = \sqrt{x}; \quad du = \frac{1}{2}x^{-\frac{1}{2}}dx$$

$$\int \frac{1}{(\sqrt{x})\sqrt{(\sqrt{2})^2 + u^2}}\left(\frac{du}{\frac{1}{2}x^{-\frac{1}{2}}}\right) = \int \frac{1}{(\sqrt{x})\sqrt{(\sqrt{2})^2 + u^2}}(2\sqrt{x})du$$

$$= \int \frac{2}{\sqrt{(\sqrt{2})^2 + u^2}}\, du = 2\sinh^{-1}\left(\frac{u}{\sqrt{2}}\right) = 2\sinh^{-1}\left(\frac{\sqrt{x}}{\sqrt{2}}\right) + C$$

Additional Problems 7.5

Given $f(x)$, find $f'(x)$.

1. $f(x) = \cosh(5x)$

2. $f(x) = \coth(x - x^{-2} + 3)$

3. $f(x) = \tanh^{-1}(x^3)$

4. $f(x) = \sinh^{-1}\left[(x+5)^3\right]$

5. $f(x) = (\cosh x)(\sinh^{-1} x)$

6. $f(x) = x^4 \sec h^{-1}(2x)$

Given $f(x)$, find $\int f(x)dx$.

7. $f(x) = \sec h^2(8x)$

8. $f(x) = \dfrac{\coth x}{\cosh x \sinh x}$

9. $f(x) = \dfrac{1}{\sqrt{x^2 - 1}}$

10. $f(x) = \dfrac{\sqrt{x}}{\sqrt{x^3 + 2}}$

7.6 Improper Integrals

As advertised in the introduction, memorization is the essential part of this chapter. We have, however, concluded the necessary memorization sections and are ready to move on to some more difficult types of calculus.

We begin by introducing integrals in which one or both of the boundaries is infinity. The official name for these integrals is *improper integrals*. In order to solve them we must take an extra step in the beginning of our integration which involves removing the infinite boundary from the integral. Once this step is taken, we just solve as usual, then take the infinite limit of the resulting function.

$$\int_a^\infty f(x)dx = \lim_{t \to \infty} \int_a^t f(x)dx$$

$$\int_{-\infty}^a f(x)dx = \lim_{t \to -\infty} \int_t^a f(x)dx$$

$$\int_{-\infty}^\infty f(x)dx = \lim_{t \to -\infty} \int_t^a f(x)dx + \lim_{t \to \infty} \int_a^t f(x)dx$$

--where a is a constant

Note: As we pointed out in Chapter 4, if a limit exists, the integral is said to be *converging*. If the limit does not exist, then the integral is said to be *diverging*.

Examples   7.6
Evaluate the following integrals.

1. $\displaystyle\int_1^\infty \frac{1}{2x+1}\,dx$

$$\int_1^\infty \frac{1}{2x+1}\,dx = \lim_{t\to\infty}\int_1^t \frac{1}{2x+1}\,dx$$

$$u = 2x+1$$

$$du = 2\,dx$$

$$\lim_{t\to\infty}\int_1^t \frac{1}{u}\left(\frac{du}{2}\right) = \lim_{t\to\infty}\left(\frac{1}{2}\ln|u|\right)$$

$$= \lim_{t\to\infty}\left(\frac{1}{2}\ln|2x+1|\right)\Big|_1^t$$

$$= \lim_{t\to\infty}\left(\frac{1}{2}\ln|2t+1| - \frac{1}{2}\ln|3|\right)$$

$$= \infty \quad \therefore \text{ diverges}$$

The most important thing about these problems is to get off on the right foot-- i.e., make sure you get that limit in there to get rid of the infinite bounds. Once you've done that, the rest of the problem is pretty straight forward.

In this particular example, we need to do a simple substitution. As you can see, this substitution leads us to make use of natural logs--something we learned about earlier in this chapter. Once the substitution is made and the integral solved, all that is left is to take the limit, something we're all masters at by now.

2. $\displaystyle\int_2^\infty \frac{1}{(2x+1)^2}\,dx$

$$\int_2^\infty \frac{1}{(2x+1)^2}\,dx = \lim_{t\to\infty}\int_2^t \frac{1}{(2x+1)^2}\,dx$$

$$u = 2x+1$$

$$du = 2\,dx$$

$$\lim_{t\to\infty}\int_2^t \frac{1}{t^2}\left(\frac{du}{2}\right) = \lim_{t\to\infty}\left[-\frac{1}{2u}\right] = \lim_{t\to\infty}\left[-\frac{1}{2(2x+1)}\right]\Big|_2^\infty$$

$$= \lim_{t\to\infty}\left[-\frac{1}{2(2t+1)} + \frac{1}{2(5)}\right]$$

$$= 0 + \frac{1}{10} = \frac{1}{10} \quad \therefore \text{ converges}$$

Similar to the last problem, we again begin by getting rid of the infinite boundary, do a simple substitution, then take the limit of the resulting function.

3. $\displaystyle\int_{-\infty}^{2}\frac{8x}{4x^2+1}dx$

This time we've got an infinite boundary that is negative. Note that this doesn't change our approach, just the bound of interest.

$$\int_{-\infty}^{2}\frac{8x}{4x^2+1}dx=\lim_{t\to-\infty}\int_{t}^{2}\frac{8x}{4x^2+1}dx$$

$$u=4x^2+1$$

$$du=8x\,dx$$

$$\lim_{t\to-\infty}\int_{t}^{2}\frac{8x}{u}\left(\frac{du}{8x}\right)=\lim_{t\to-\infty}\int_{t}^{2}\frac{1}{u}du$$

$$=\lim_{t\to-\infty}\ln|u|=\lim_{t\to-\infty}\ln\left|4x^2+1\right|\Big|_{t}^{2}=\infty\ \therefore\ \text{diverges}$$

4. $\displaystyle\int_{-\infty}^{\infty}\frac{2}{5x^5}dx$

Of special note in this problem is the fact that both bounds are infinite. Hence, when we set up the integrals with the limits, we need to pick a constant as one of the bounds for both integrals. Usually you pick something like 1 or 0. Here, we picked 1 since 0 causes the functions to not exist.

$$\int_{-\infty}^{\infty}\frac{2}{5x^5}dx=\lim_{t\to-\infty}\int_{t}^{1}\frac{2}{5x^5}dx+\lim_{t\to\infty}\int_{1}^{t}\frac{2}{5x^5}dx$$

$$=\lim_{t\to-\infty}\int_{t}^{1}\frac{2}{5}x^{-5}dx+\lim_{t\to\infty}\int_{1}^{t}\frac{2}{5}x^{-5}dx$$

$$=\lim_{t\to-\infty}\left[\frac{2}{5}\left(-\frac{1}{4}x^{-4}\right)\right]_{t}^{1}+\lim_{t\to\infty}\left[\frac{2}{5}\left(-\frac{1}{4}x^{-4}\right)\right]_{1}^{t}$$

$$=\lim_{t\to-\infty}\left[-\frac{1}{10}x^{-4}\right]_{t}^{1}+\lim_{t\to\infty}\left[-\frac{1}{10}x^{-4}\right]_{1}^{t}$$

$$=-\frac{1}{10}\left(\frac{1}{(1)^4}\right)-\left(-\frac{1}{10}\right)\left(\frac{1}{-\infty}\right)+\left(-\frac{1}{10}\right)\left(\frac{1}{\infty}\right)-\left(-\frac{1}{10}\right)\left(\frac{1}{(1)^4}\right)$$

$$=-\frac{1}{10}-0+0+\frac{1}{10}=0\ \therefore\ \text{converges}$$

5. $\displaystyle\int_{-\infty}^{\infty}\frac{1}{1+x^2}dx$

In this problem, again, both bounds are infinite. Fortunately, since 0 doesn't cause the function to not exist, so we can use 0 as our constant. This simplifies things quite a bit.

$$\int_{-\infty}^{\infty}\frac{1}{1+x^2}dx=\lim_{t\to-\infty}\int_{t}^{0}\frac{1}{1+x^2}dx+\lim_{t\to\infty}\int_{0}^{t}\frac{1}{1+x^2}dx$$

$$=\lim_{t\to-\infty}\left(\tan^{-1}x\right)\Big|_{t}^{0}+\lim_{t\to\infty}\left(\tan^{-1}x\right)\Big|_{0}^{t}$$

$$= \tan^{-1} 0 - \tan^{-1}(-\infty) + \tan^{-1}(\infty) - \tan^{-1}(0)$$

$$= 0 - \left(-\frac{\pi}{2}\right) + \left(\frac{\pi}{2}\right) - 0 = \pi \quad \therefore \text{ converges}$$

Additional Problems 7.6

Evaluate the following integrals.

1. $\int_{1}^{\infty} \frac{1}{\sqrt{x}} dx$

4. $\int_{-\infty}^{\infty} \cos x \, dx$

2. $\int_{-\infty}^{2} \frac{1}{\left(1 - x^2\right)} dx$

5. $\int_{-\infty}^{\infty} \sec hx \, dx$

3. $\int_{-\infty}^{\infty} \frac{1}{5 + x} dx$

7.7 Integrals Operating on Discontinuous Functions

Similar to improper integrals in terms of how we solve them, we now move on to the study of integrals that operate on discontinuous functions over the interval in which the discontinuity occurs. For example, let's say we want to integrate a function from 1 to 5, but the function is discontinuous at 2. We cannot simply integrate as is or we'll get a nonsense answer. Instead, we have to make some adjustments, and as in the last section these involve limits.

Method for solving discontinuous functions:

1. Divide the integrals at the point of discontinuity. For example, given $\int_{1}^{5} f(x)dx$, where $f(x)$ is discontinuous at 2, we divide the integrals as follows:

$$\int_{1}^{2} f(x)dx + \int_{2}^{5} f(x)dx$$

2. We still cannot integrate because one of the boundaries in each of the integrals is discontinuous in the function. So we make the following modification using one-sided limits:

$$\lim_{t \to 2^-} \int_{1}^{2} f(x)dx + \lim_{t \to 2^+} \int_{2}^{5} f(x)dx$$

3. Solve both integrals, and then take the limits of the resulting functions. If there is a limit in both cases, the integral converges. If one or more of the limits does not exist, however, the integral diverges.

Hurricane Calculus

Sometimes we'll get lucky and the problem will start out with the discontinuity as one of the boundary conditions. In these cases we skip Step #1 and proceed to Step #2.

Examples   **7.7**

Evaluate the following integrals.

1. $\int_0^2 \frac{1}{x} dx$

$\int_0^2 \frac{1}{x} dx = \lim_{t \to 0^+} \int_t^2 \frac{1}{x} dx = \lim_{t \to 0^+} \left(\ln|x| \right)\Big|_t^2$

$= \lim_{t \to 0^+} \left(\ln|2| - \ln|t| \right) = \infty \quad \therefore \text{ diverges}$

In this problem we are lucky because the point of discontinuity is already the lower bound. Hence, all we have to do is throw on the limit, modifying the bound, and solve.

2. $\int_0^1 \frac{1}{(x-1)^{\frac{2}{3}}} dx$

$\int_0^1 \frac{1}{(x-1)^{\frac{2}{3}}} dx = \lim_{t \to 1^-} \int_0^t \frac{1}{(x-1)^{\frac{2}{3}}} dx$

$u = x - 1$

$du = dx$

$\lim_{t \to 1^-} \int_0^t \frac{1}{u^{\frac{2}{3}}} dx = \lim_{t \to 1^-} \left(3u^{\frac{1}{3}} \right) = \lim_{t \to 1^-} \left(3(x-1)^{\frac{1}{3}} \right)\Big|_0^t$

$= 3\left[(t-1)^{\frac{1}{3}} - (0-1)^{\frac{1}{3}} \right]$

$= 3 \quad \therefore \text{ converges}$

Again in this problem, one of the bounds contains the point of discontinuity, so all we have to do is throw on the limits and go. In this case, we need to do a simple substitution to solve.

5. $\int_0^5 \frac{1}{(2-x)^{\frac{1}{3}}} dx$

$\int_0^5 \frac{1}{(2-x)^{\frac{1}{3}}} dx = \lim_{t \to 2^-} \int_0^2 \frac{1}{(2-x)^{\frac{1}{3}}} dx + \lim_{t \to 2^+} \int_2^5 \frac{1}{(2-x)^{\frac{1}{3}}} dx$

$u = 2 - x$

$du = dx$

Here we have a function in which the point of discontinuity is within the interval of integration. So, we have to break the interval at the point of discontinuity before we begin.

$$\lim_{t \to 2^-} \int_0^2 \frac{1}{u^{\frac{1}{3}}} dx + \lim_{t \to 2^+} \int_2^5 \frac{1}{u^{\frac{1}{3}}} dx = \lim_{t \to 2^-} \left(\frac{3}{2} u^{\frac{2}{3}} \right) + \lim_{t \to 2^+} \left(\frac{3}{2} u^{\frac{2}{3}} \right)$$

$$= \lim_{t \to 2^-} \left(\frac{3}{2}(2-x)^{\frac{2}{3}} \right) \Big|_0^t + \lim_{t \to 2^+} \left(\frac{3}{2} u^{\frac{2}{3}} \right) \Big|_t^5$$

$$= \lim_{t \to 2^-} \left(\frac{3}{2} \left[(2-t)^{\frac{2}{3}} - (2-0)^{\frac{2}{3}} \right] \right) + \lim_{t \to 2^+} \left(\frac{3}{2} \left[(2-5)^{\frac{2}{3}} - (2-t)^{\frac{2}{3}} \right] \right)$$

$$= 0 - \frac{3}{2}(2)^{\frac{2}{3}} + \frac{3}{2}(-3)^{\frac{2}{3}} - 0 \quad \therefore \text{ converges}$$

Additional Problems 7.7

Evaluate the following integrals.

1. $\int_{-1}^2 \frac{1}{x+1} dx$

2. $\int_0^4 \frac{1}{\sqrt{4-x}} dx$

3. $\int_{-2}^2 \frac{1}{x^2} dx$

4. $\int_{-1}^1 \frac{1}{x} dx$

5. $\int_{-3}^0 \frac{1}{(1+x)^3} dx$

Chapter 8

Techniques for Solving Mind-Bending Integrals

This is it! The last of the four integral chapters. Of course, we're not going to be getting off lightly here. In this chapter we learn how to solve the really difficult integrals. It's not particularly pleasant, but having gotten through Chapters 5, 6 and 7, we're ready. In order to succeed, we need to commit the step-by-step problem-solving procedures to memory and work a lot of examples. Fortunately, this is not hard.

We're about to encounter some really nasty integrals in this chapter. In fact, we're about to encounter the hardest integrals in first year calculus, period. If you look over the heading for this chapter, you'll notice that these integrals come in six basic varieties in terms of how we're going to go about solving them. In all, there are actually seven new techniques we're going to have to learn. This sounds like a lot, but actually it's not too bad since we've gone to great lengths to ensure that all of the steps are laid out for you in meticulous detail. With enough work on the example problems, you should find that you can solve these integrals without too much difficulty. Be warned, however: You're going to have to work a lot of examples before you're comfortable with this material. Please take some extra time to study all of the examples provided in this chapter. It will be time well invested.

8.1 Integration by Parts

The first of the seven techniques that we must learn in this chapter is *integration by parts*. This technique is powerful and easy to use. It's based on the following equation:

$$\int u\,dv = uv - \int v\,du$$

So what is this equation saying? It's saying that if you have an integral operating on a really nasty function, and that function can readily be broken down into two functions (here we're calling them u and dv), then we can rewrite the integral as the function u times the function v (v is just the antiderivative of dv) minus the integral of the functions v times du (du is just the derivative of u). Admittedly, this sounds kind of complicated, but when you see it actually worked out in some examples it's not that bad.

As you have probably guessed, the new integral created via the above equation is suppose to be easier to solve than the original. If you break the original function up in a wise manner, this is usually true. Hence, the trick here is to learn how to do this little transformation wisely. The examples that follow illustrate.

Examples   **8.1**

Evaluate the following integrals.

1. $\int xe^{-x}dx$

    $u = x$ and $dv = e^{-x}dx$

    $du = dx$ and $v = -e^{-x}$

    $\therefore \int xe^{-x}dx = -xe^{-x} - \int(-e^{-x})dx$

    $\qquad = -xe^{-x} + \int e^{-x}dx = -xe^{-x} - e^{-x} + C$

To use integration by parts on this integral we first have to figure out how to break the function into two parts. Remember, the idea here is to get a simpler integral. Generally, the way we accomplish this is by picking the u to be something that, after taking the derivative, is nicer, while the dv is something that upon taking the antiderivative won't be that bad.

2. $\int x^2 e^x dx$

    $u = x^2$ and $dv = e^x dx$

    $du = 2x\, dx$ and $v = e^x$

    $\therefore \int x^2 e^x dx = x^2 e^x - \int e^x(2x)dx$

    $\qquad = x^2 e^x - 2\int xe^x dx$

    Note: We still can't solve this integral, so we have to do a second integration by parts on the new integral.

    $u = x$ and $dv = e^x dx$

    $du = dx$ and $v = e^x$

    $\int xe^x dx = xe^x - \int e^x dx$

    $\qquad xe^x - e^x + C$

    $\therefore \int x^2 e^x dx = x^2 e^x - 2[xe^x - e^x] + C$

This example is a bit more involved than the first one simply because the function being operated on by the integral is a lot nastier.

Again, we begin by splitting the original function into two parts. Remember, we want the u to be the part that can be broken down easily via a derivative, and the dv to be the part that won't get real nasty upon taking its derivative. Clearly, in this case, the x squared term is the obvious choice for u since it can be broken down easily.

Upon completing the integration by parts, however, we find that the new integral still can't be solved. Hence, we have to do a second integration by parts.

Once the second integration by parts is complete, we must add the results of both to get the answer.

3. $\int x\cos 3x\,dx$

In this problem integration by parts enables us to create a new integral that is readily solvable via the old substitution we learned back in Chapter 5.

$u = x$ and $dv = \cos 3x\,dx$

$du = dx$ and $v = \dfrac{1}{3}\sin 3x$

As you can see, when doing integration by parts we're going to end up using other techniques in conjunction with it. Don't be fooled.

$\int x\cos 3x\,dx = x\left(\dfrac{1}{3}\sin 3x\right) - \int\left(\dfrac{1}{3}\sin 3x\right)dx$

$u = 3x$

$du = 3dx$

$\int\left(\dfrac{1}{3}\sin u\right)\left(\dfrac{du}{3}\right) = \dfrac{1}{9}(-\cos u) = -\dfrac{1}{9}\cos 3x$

After doing the standard substitution, we just add the results with those from the integration by parts.

$\therefore\ \int x\cos 3x\,dx = x\left(\dfrac{1}{3}\sin 3x\right) - \left[-\dfrac{1}{9}\cos 3x\right] + C$

4. $\int \tan^{-1} x\,dx$

The natural impulse upon seeing this problem is to flip back through Chapter 7 and try to find out what the integral of arc tangent of x is. You will, upon doing this, however, note that we didn't define it there, hence we must solve the hard way.

$u = \tan^{-1} x$ and $dv = dx$

$du = \dfrac{1}{1+x^2}\,dx$ and $v = x$

In doing so, we find that this problem is very similar to the previous one in that after doing the integration by parts, normal substitution is required to solve the resulting integral.

$\int \tan^{-1} x\,dx = \left(\tan^{-1} x\right)(x) - \int x\left(\dfrac{1}{1+x^2}\right)dx$

$u = 1 + x^2$

$du = 2x\,dx$

$\int x\left(\dfrac{1}{u}\right)\left(\dfrac{du}{2x}\right) = \dfrac{1}{2}\ln|u| = \dfrac{1}{2}\ln\left|1 + x^2\right| + C$

As before, after solving the integral, its results must be added to the rest of the integration by parts results to get the answer.

$\therefore\ \int \tan^{-1} x\,dx = \left(\tan^{-1} x\right)(x) - \dfrac{1}{2}\ln\left|1 + x^2\right| + C$

5. $\int \dfrac{\ln x}{x^2}\,dx$

There's nothing really out of the ordinary about this problem, we show it to you only because the way we break the original function up may not be obvious at first glance.

$u = \ln x$ and $dv = x^{-2}dx$

$du = \dfrac{1}{x}\,dx$ and $v = -x^{-1}$

$\int \dfrac{\ln x}{x^2}\,dx = (\ln x)\left(-\dfrac{1}{x}\right) - \int\left(-\dfrac{1}{x}\right)\left(\dfrac{1}{x}\right)dx = -\dfrac{\ln x}{x} + \int\dfrac{1}{x^2}\,dx = -\dfrac{\ln x}{x} + \left(-\dfrac{1}{x}\right) + C$

Additional Problems 8.1

Evaluate the following integrals.

1. $\int xe^x dx$

2. $\int x^2 \sin x \, dx$

3. $\int \cos(\ln x) dx$

4. $\int (\ln x)^2 dx$

5. $\int x \tan^{-1} x \, dx$

6. $\int \sin^2 x \, dx$

7. $\int \dfrac{x^3}{\sqrt{x^2+1}} dx$

8.2 Trig Integral Tricks

Our second technique is less specific than integration by parts. In point of fact, there is a great deal of "hand-waving" that goes on to even call this bag of tricks a technique. Nevertheless, you will find it useful for some of the more complex trig integrals.

The basic idea behind this second technique is to take a complex trig integral you cannot solve right off, massage it a bit, and then do a standard *u* substitution to solve. The key here is the massage. And the key to the massage is knowledge of the basic trigonometric identities--so brush up on them now if you're rusty (see Chapter 1, section 1.5).

There are three main types of complex trig integrals that you'll be able to use this technique on:

> Type 1: Integrals in the form: $\int \sin^n x \, dx$, $\int \cos^n x \, dx$
>
> Type 2: Integrals in the form: $\int \cos^n x \sin^m x \, dx$
>
> Type 3: Integrals in the form: $\int \tan^n x \sec^m x \, dx$
>
> --where *n* and *m* are positive integers.

When you see an integral in one of the above forms, little bells should go off telling you its time to use trig integral tricks. In order to ensure that these little bells go off, make sure you commit all three types to memory, then study the following steps for solving each.

Methods for Solving

Type 1: Integrals in the form: $\int \sin^n x \, dx$, $\int \cos^n x \, dx$

1. If n is *even*, convert using the appropriate half-angle formula and solve using a standard u substitution. The two half-angle formulas of concern are:

$$\sin^2 x = \frac{1 - \cos 2x}{2} \quad \text{and} \quad \cos^2 x = \frac{1 + \cos 2x}{2}$$

2. If n is *odd*, convert as follows:

$$\int \sin^n x \, dx = \int \left(\sin^{n-1} x\right)\left(\sin x\right) dx \quad \text{or} \quad \int \cos^n x \, dx = \int \left(\cos^{n-1} x\right)\left(\cos x\right) dx$$

Once this is done, convert the $n - 1$ term (which is now even), using the formula, $\cos^2 x + \sin^2 x = 1$.

For example, if the $n - 1$ term was $\cos^6 x$, we would convert it to $\left(\cos^2 x\right)^3$. Having done this, we would then convert it to, $\left(1 - \sin^2 x\right)^3$. After multiplying everything out in the integral and doing the usual substitution, we find that the term that was originally factored out is cancelled out, and the problem readily solved.

Type 2: Integrals in the form: $\int \cos^n x \sin^m x \, dx$

1. If n and m are even, then convert both using half - angle formulas and solve.

2. If n is odd then convert as follows:

$$\int \cos^n x \sin^m x \, dx = \int \cos^{n-1} x \cos x \sin^m x \, dx$$

Convert the $n - 1$ term using $\cos^2 x + \sin^2 x = 1$ and solve with a normal u substitution letting $u = \sin x$.

3. If m is odd, then convert as follows:

$$\int \cos^n x \sin^m x \, dx = \int \cos^n x \sin^{m-1} x \sin x \, dx$$

Convert the $m - 1$ term using $\cos^2 x + \sin^2 x = 1$ and solve with a normal u substitution letting $u = \cos x$

Type 3: Integrals in the form: $\int \tan^n x \sec^m x \, dx$

1. If n is even and m is odd, then back up 15 and punt on this method immediately. Another method is required to solve this type of integral.

2. If n is odd then convert as follows:

$$\int \tan^n x \sec^m x \, dx = \int \tan^{n-1} x \sec^{m-1} x \tan x \sec x \, dx$$

Convert the n - 1 term to secant using $\tan^2 x = \sec^2 x - 1$ and solve with a normal u substitution letting $u = \sec x$.

3. If m is even, then convert as follows:

$$\int \tan^n x \sec^m x \, dx = \int \tan^n x \sec^{m-2} x \sec^2 x \, dx$$

Convert the m - 2 term using $\tan^2 x = \sec^2 x - 1$ and solve with a normal u substitution letting $u = \tan x$

Examples 8.2

Evaluate the following integrals.

1. $\int \cos^4 x \, dx$

In solving this problem, we first recognize that it is a clear Type #1 problem. So, we follow the procedures outlined for Type #1s.

$$\int \cos^4 x \, dx = \int \left(\cos^2 x \right)^2 dx$$

$$= \int \left(\frac{1 + \cos 2x}{2} \right)^2 dx$$

$$= \int \frac{1 + 2\cos 2x + \cos^2 2x}{4} dx$$

$$= \int \frac{1}{4} dx + \int \frac{1}{2} \cos 2x \, dx + \int \frac{1}{4} \cos^2 2x \, dx$$
$$\quad (1) \qquad\qquad (2) \qquad\qquad\quad (3)$$

Since n is even, (4), we simply break it down as shown to the left, then substitute in the equivalent half-angle formula.

Note that upon doing this we end up getting three different integrals. Each one needs to be solved separately, so we designate each one as 1, 2, and 3 respectively.

(1) $\int \frac{1}{4} dx = \frac{1}{4} x$

(2) $\int \frac{1}{2} \cos 2x \, dx$

$$u = 2x$$
$$du = 2 \, dx$$

$$\int \frac{1}{2} \cos u \left(\frac{du}{2} \right) = \frac{1}{4} \sin u = \frac{1}{4} \sin 2x$$

Solving the first two integrals is relatively simple. The first one is a no-brainer if you've been paying any attention at all the during the last three chapters. The second one requires a simple substitution..

(3) $\int \frac{1}{4}\cos^2 2x \, dx = \frac{1}{4}\int\left(\frac{1+\cos 2x}{2}\right)dx$

$$= \frac{1}{8}\int(1+\cos 2x)dx$$

$$= \frac{1}{8}x + \frac{1}{8}\left(\frac{1}{2}\sin 2x\right)$$

The third integral is harder to solve. It requires that we use the same technique we started with, because it's Type #1.

$\therefore \quad \int \cos^4 x \, dx = \frac{1}{4}x + \frac{1}{4}\sin 2x + \left[\frac{1}{8}x + \frac{1}{16}\sin 2x\right]$

            (1)   (2)   (3)

After we solve all three integrals, we add their results together to come up with our final answer.

2. $\int \sin^7 x \, dx$

Here we have another Type #1 integral, but this time n is odd, so we have to proceed to Step #2 to solve. This isn't so bad, following along with the rules. The most difficult thing is multiplying out the cubed term after making switch from sine squared to 1-cosine squared.

$\int \sin^7 x \, dx = \int\left(\sin^2 x\right)^3 \sin x \, dx$

$$= \int\left(1-\cos^2 x\right)^3 \sin x \, dx$$

$$= \int\left(1-3\cos^2 x + 3\cos^4 x - \cos^6 x\right)\sin x \, dx$$

$$u = \cos x$$

$$du = -\sin x \, dx$$

$$= \int\left(1-3u^2 + 3u^4 - u^6\right)\sin x \left(\frac{du}{-\sin x}\right)$$

$$= -\int\left(1-3u^2 + 3u^4 - u^6\right)du$$

$$= -\left[u - u^3 + \frac{3}{5}u^5 - \frac{1}{7}u^7\right]$$

$$= -\cos x + \cos^3 x - \frac{3}{5}\cos^5 x + \frac{1}{7}\cos^7 x + C$$

After doing the multiplication, the problem is easy. We just do a simple u substitution, and solve the resulting elementary integrals.

3. $\int \cos^3 x \sin^2 x \, dx$

This integral is a Type #2, and a pretty easy one at that. We just follow the guidelines as outlined on page 166, and end up doing a simple u substitution.

$\int \cos^3 x \sin^2 x \, dx = \int \cos^2 x \sin^2 x \cos x \, dx$

$$= \int\left(1-\sin^2 x\right)\sin^2 x \cos x \, dx$$

$$u = \sin x, \; du = \cos x \, dx$$

$$= \int (1 - u^2) u^2 \cos x \left(\frac{du}{\cos x} \right)$$

$$= \int (u^2 - u^4) du = \frac{1}{3} u^3 - \frac{1}{5} u^5 = \frac{1}{3} \sin^3 x - \frac{1}{5} \sin^5 x + C$$

4. $\int \cos^2 x \sin^2 x \, dx$

This problem starts out very straightforward. We've got a classic Type #2 case, so we proceed to solve as outlined above. When we do this, however, we end up with another integral that is of the Type #1 variety.

$$\int \cos^2 x \sin^2 x \, dx = \int \left(\frac{1 + \cos 2x}{2} \right) \left(\frac{1 - \cos 2x}{2} \right) dx$$

$$= \frac{1}{4} \int (1 - \cos^2 2x) dx$$

$$= \frac{1}{4} x - \frac{1}{4} \int \cos^2 2x \, dx$$

$-\frac{1}{4} \int \cos^2 2x \, dx$, is a Type #1, so we must solve it

using Type #1 techniques.

One critical point to note here in solving a Type #1:

Due to the fact that the cosine squared term is operating on 2x, when we convert to the half-angle formula it becomes a 4x. See the formulas above if you need further clarification.

$$-\frac{1}{4} \int \cos^2 2x \, dx = -\frac{1}{4} \int \left(\frac{1 + \cos 4x}{2} \right) dx$$

$$= -\frac{1}{8} \int (1 + \cos 4x) dx$$

$$= -\frac{1}{8} x - \frac{1}{8} \left(\frac{1}{4} \sin 4x \right)$$

$$\int \cos^2 x \sin^2 x \, dx = \frac{1}{4} x - \frac{1}{8} x - \frac{1}{8} \left(\frac{1}{4} \sin 4x \right) + C$$

5. $\int \tan^3 x \sec^2 x \, dx$

Finally, we end our run of examples here with an example of a Type #3 integral. These are easy to identify, and usually easy to solve. All you need to do is follow the basic steps. Usually we don't find any additional tricks with Type #3 integrals.

$$\int \tan^3 x \sec^2 x \, dx = \int \tan^2 x \sec x \tan x \sec x \, dx$$

$$= \int (\sec^2 x - 1) \sec x \tan x \sec x \, dx$$

$$u = \sec x$$

$$du = \sec x \tan x \, dx$$

$$= \int \left(u^2 - 1\right) u \tan x \sec x \left(\frac{du}{\sec x \tan x}\right)$$

$$= \int \left(u^3 - u\right) du$$

$$= \frac{1}{4}u^4 - \frac{1}{2}u^2 = \frac{1}{4}(\sec x)^4 - \frac{1}{2}(\sec x)^2 + C$$

Additional Problems　　　　　　　　　　　　　　　　　　**8.2**

Evaluate the following integrals.

1. $\int \sin^4 x\, dx$　　　　　　　　　3. $\int \tan^6 x\, dx$

2. $\int \sin^5 x \cos^3 x\, dx$　　　　4. Sometimes you'll get problems requiring knowledge of trig product such as the following: $\int \cos 4x \cos 3x\, dx$

8.3 Trig Substitution

As long as we're on the subject of trig, we might as well learn the final integration technique that specifically uses it, trig substitution. Though its name would seem to imply that we're going to use it to attack certain complex trig integrals, in fact we're going to use it to deal with integrals containing one of three types of the following radicals:

$$\sqrt{a^2 - x^2},\ \sqrt{a^2 + x^2},\ \sqrt{x^2 - a^2}$$

If you see any of these expressions in an integral, you may be sure trig substitution is the technique of choice. Here's what you do if and when you see them:

1. Substitute the appropriate trig function in for x:

    a) When you see $\sqrt{a^2 - x^2}$, substitute $a \sin\theta$ in for x.

    b) When you see $\sqrt{a^2 + x^2}$, substitute $a \tan\theta$ in for x.

    c) When you see $\sqrt{x^2 - a^2}$, substitute $a \sec\theta$ in for x.

2. You'll notice that after making one of these substitutions we'll be able to take advantage of an appropriate trig identity and wipe out the radical. Do this, and then solve the integral--a relatively easy task now.

3. After solving we have an answer, but it's not in terms of x, so substitute x back in. How? Use geometry. Draw a right triangle with the values from the original substitution. Using this triangle, convert back to x using the basic geometric relationships of sin, cos, tan, etc.

For example, see Fig. 8.1. If we substituted $a\sin\theta$ for x in the beginning of the problem, we would set up the triangle with θ as an angle, x opposite it, and a as the hypotenuse (because sine equals opposite over hypotenuse). The other side of the triangle, the one adjacent to θ, must be solved for using the Pythagorean theorem. If one the terms in our answer was $(5a\cos\theta)$, then converting back, we would get:

$5a\left(\dfrac{\sqrt{a^2-x^2}}{a}\right)$. How did we know $\cos\theta$ was equal to $\left(\dfrac{\sqrt{a^2-x^2}}{a}\right)$?

Fig. 8.1

Because cosine is equal to adjacent over the hypotenuse--the adjacent side of the triangle is $\sqrt{a^2-x^2}$, the hypotenuse is a.

Summarizing the technique:
1. Substitute the appropriate trig function in for x.
2. Solve the integral taking advantage of the appropriate trig identities.
3. Convert the answer back to x by taking advantage of the basic geometric relationships of a right triangle whose characteristics come from the original substitution.

Examples   8.3

Evaluate the following integrals.

1. $\displaystyle\int \frac{1}{x^2\sqrt{25-x^2}}\,dx$

    i. $\sqrt{(5)^2-x^2}$ $\therefore$ substitute $x=5\sin\theta$, $dx=5\cos\theta\,d\theta$

$$\int\frac{1}{x^2\sqrt{25-x^2}}\,dx=\int\frac{1}{(5\sin\theta)^2\sqrt{25-(5\sin\theta)^2}}(5\cos\theta\,d\theta)$$

    ii.
$$=\int\frac{5\cos\theta}{(5\sin\theta)^2\,5\sqrt{1-\sin^2\theta}}\,d\theta$$

$$=\int\frac{5\cos\theta}{(25\sin^2\theta)5\sqrt{\cos^2\theta}}\,d\theta$$

$$=\int\frac{\cos\theta}{(25\sin^2\theta)(\cos\theta)}\,d\theta$$

$$=\int\frac{1}{25\sin^2\theta}\,d\theta=\frac{1}{25}\int\csc^2\theta\,d\theta=-\frac{1}{25}\cot\theta+C$$

To solve this problem we just go through the 3 steps outlined above. Once you identify what the initial substitution should be, the problem is fairly standard until you get to part 3 where you convert back.

As you can see, solving this integral is not too bad. About the most serious thing you have to contend with is some algebra.

iii. Convert back to x, $x = 5\sin\theta$ $\therefore$ $\sin\theta = \dfrac{x}{5}$

This is the crucial step. We're converting our answer back to terms of x. To do this, we draw a right triangle with sides and angle corresponding to our initial substitution. To do this, you need to remember your basic definitions of sine, cosine, etc. Once the triangle is drawn, we look at it to convert terms in our answer. In this case, we have cot. Cot, remember, is equal to adjacent over opposite, hence we look at the triangle and plug away. See Chapter 2, section 1.5 for a review if necessary.

So, $-\dfrac{1}{25}\cot\theta + C = -\dfrac{1}{25}\left(\dfrac{\sqrt{25-x^2}}{x}\right) + C$

2. $\displaystyle\int \dfrac{x^3}{\sqrt{x^2+5}}\,dx$

Our initial substitution in this problem is fairly easy to recognize. However, solving the resulting integral is more challenging than in the previous example.

i. $\sqrt{x^2 + \left(\sqrt{5}\right)^2}$, $\therefore$ Substitute $x = \sqrt{5}\tan\theta$, $dx = \sqrt{5}\sec^2\theta\,d\theta$

$\displaystyle\int\dfrac{x^3}{\sqrt{x^2+5}}\,dx = \int\dfrac{\left(\sqrt{5}\tan\theta\right)^3}{\sqrt{\left(\sqrt{5}\tan\theta\right)^2+5}}\left(\sqrt{5}\sec^2\theta\,d\theta\right)$

ii. $\displaystyle = \int\dfrac{5\sqrt{5}\tan^3\theta}{\sqrt{\left(\sqrt{5}\right)^2\tan^2\theta+\left(\sqrt{5}\right)^2}}\left(\sqrt{5}\sec^2\theta\,d\theta\right)$

In solving the integral that results after making our initial substitution, we have two major difficulties to overcome:

$\displaystyle = \int\dfrac{5\sqrt{5}\tan^3\theta}{\left(\sqrt{5}\right)\sqrt{\tan^2\theta+1}}\left(\sqrt{5}\sec^2\theta\,d\theta\right)$

1. There's a lot of algebra required to get rid of the radical.

$\displaystyle = \int\dfrac{5\sqrt{5}\tan^3\theta\sec^2\theta}{\sqrt{\sec^2\theta}}\,d\theta$

2. We have to use Trig Integral tricks, specifically Type #3, to solve.

$\displaystyle = 5\sqrt{5}\int\tan^3\theta\sec\theta\,d\theta$

$\displaystyle = 5\sqrt{5}\int\tan^2\theta\sec\theta\tan\theta\,d\theta$

$\displaystyle = 5\sqrt{5}\int\left(\sec^2\theta-1\right)\sec\theta\tan\theta\,d\theta$

$u = \sec\theta$, $du = \sec\theta\tan\theta\,d\theta$

$$= 5\sqrt{5}\int\left(u^2 - 1\right)\sec\theta\,\tan\theta\left(\frac{du}{\sec\theta\,\tan\theta}\right)$$

$$= 5\sqrt{5}\int\left(u^2 - 1\right)du$$

$$= 5\sqrt{5}\left(\frac{1}{3}u^3 - u\right) = 5\sqrt{5}\left(\frac{1}{3}(\sec\theta)^3 - (\sec\theta)\right) + C$$

iii. Convert back to x. $x = \sqrt{5}\tan\theta$, $\therefore$ $\tan\theta = \dfrac{x}{\sqrt{5}}$

As in the last problem, the final step is converting back to terms of x. Here, x is expressed in terms of tangent, so we define our triangle based on this. Then, we look at our answer, which has secant (hypotenuse over adjacent), and plug away.

$$\therefore\; 5\sqrt{5}\left(\frac{1}{3}(\sec\theta)^3 - (\sec\theta)\right) + C = 5\sqrt{5}\left(\frac{1}{3}\left(\frac{\sqrt{x^2+5}}{\sqrt{5}}\right)^3 - \left(\frac{\sqrt{x^2+5}}{\sqrt{5}}\right)\right) + C$$

3. $\displaystyle\int \frac{1}{x^2\sqrt{x^2-25}}\,dx$

Our final example involving trig substitution is very straight forward. Besides the algebra, which all these problems require to get rid of the radical, there's really nothing to this. Even the conversion back to x at the end is simple.

i. $\sqrt{x^2 - 25}$, $\therefore$ Substitute $x = 5\sec\theta$, $dx = 5\sec\theta\,\tan\theta\,d\theta$

$$\int \frac{1}{x^2\sqrt{x^2-25}}\,dx = \int \frac{1}{\left(5\sec\theta\right)^2\sqrt{\left(5\sec\theta\right)^2 - 25}}\left(5\sec\theta\,\tan\theta\,d\theta\right)$$

ii.

$$= \int \frac{5\sec\theta\,\tan\theta}{\left(25\sec^2\theta\right)\sqrt{\left(5\sec\theta\right)^2 - 25}}\,d\theta$$

$$= \int \frac{\tan\theta}{\left(5\sec\theta\right)5\sqrt{\tan^2\theta}}\,d\theta$$

$$= \frac{1}{25}\int \frac{1}{\sec\theta}\,d\theta = \frac{1}{25}\int\cos\theta\,d\theta = \frac{1}{25}\sin\theta + C$$

iii. Convert back to x: $x = 5\sec\theta$, $\therefore$ $\sec\theta = \dfrac{x}{5}$

$$\therefore\; \frac{1}{25}\sin\theta + C = \frac{1}{25}\left(\frac{\sqrt{x^2-25}}{x}\right) + C$$

Additional Problems   **8.3**

Evaluate the following integrals.

1. $\int \dfrac{1}{x\sqrt{36-x^2}}\,dx$

3. $\int \dfrac{1}{\sqrt{(x^2-1)^3}}\,dx$

5. $\int \dfrac{1}{(25+x^2)^2}\,dx$

2. $\int \dfrac{\sqrt{9+x^2}}{x}\,dx$

4. $\int \left(\sqrt{36-4x^2}\right)dx$

8.4 Completing the Square

"Oh no, they're back!" you shriek. "Demons from Algebra II!" Well, yes. Unfortun‑tely for all you algebra phobes out there, we're going to be faced with some integrals in this chapter that operate on functions containing quadratics that we just can't integrate without using the old Algebra II, completing the square technique.

You remember completing the square, the idea being to take a non-factorable quadratic equation and make it factorable. Why would we want to do this here? Well, a factorable quadratic lends itself quite nicely to easy substitutions and hence simple integration.

As you will recall, completing the square is done as follows:

Given the nonfactorable quadratic, $ax^2 + bx + c$, we can rewrite it as:

$$ax^2 + bx + c = a\left[x + \left(\frac{b}{2a}\right)\right]^2 + c - \frac{b^2}{4a}$$

Examples   **8.4**

Complete the square.

1. $x^2 - 6x + 17$

$$= (1)\left[x + \frac{(-6)}{2(1)}\right]^2 + 17 - \frac{(-6)^2}{4(1)} = (x-3)^2 + 8$$

As you can see, all we're doing here is following the above formula to the letter. We're merely translating the given quadratics into a new form.

2. $8x^2 + 13x - 1$

$$= (8)\left[x + \frac{13}{2(8)}\right]^2 + (-1) - \frac{(13)^2}{4(8)} = 8\left(x + \frac{13}{16}\right)^2 - \frac{201}{32}$$

When we run into these quadratics in integrals, having them in this new form is much easier to deal with.

Now that we're all brushed up on the technique of completing the square, we can move on and use it to solve some integrals. Remember, this technique is just going to be used when we see an integral with a nasty quadratic in it. When we do, we simply:

1. Complete the square.
2. Make a standard substitution letting u equal the inside of the perfect square.
3. Solve.

Examples 8.45

Evaluate the following integrals.

1. $\int \dfrac{1}{x^2 - 6x + 17}\, dx$

Very straight forward problem. We see that we've got a nasty quadratic in the denominator, we translate it via completing the square into an easier form, then do a simple u substitution and we're done.

i. $x^2 - 6x + 17 = 1\left[x + \dfrac{(-6)}{2(1)} \right]^2 + 17 - \dfrac{36}{4(1)} = (x - 3)^2 + 8$

ii. $\int \dfrac{1}{(x-3)^2 + 8}\, dx$

$$u = x - 3$$
$$du = dx$$

$$= \int \dfrac{1}{u^2 + 8}\, du$$

$$= \int \dfrac{1}{u^2 + \left(\sqrt{8}\right)^2}\, du = \dfrac{1}{2\sqrt{2}} \tan^{-1}\left(\dfrac{u}{2\sqrt{2}}\right) = \dfrac{1}{2\sqrt{2}} \tan^{-1}\left(\dfrac{x-3}{2\sqrt{2}}\right) + C$$

2. $\int \dfrac{19 + 96x}{x^2 - 2x + 2}\, dx$

This problem begins much like the last one. We simply complete the square and get ready to solve.

i. $x^2 - 2x + 2 = 1\left[x + \dfrac{(-2)}{2(1)} \right]^2 + 2 - \dfrac{(-2)^2}{4(1)} = (x - 1)^2 + 1$

Important point: As you are about to see, the real difficulty with these problems comes in solving the altered integral. Completing the square is the easy part.

ii. $\int \dfrac{19 + 96x}{(x-1)^2 + 1}\, dx$

$$u = x - 1, \quad \therefore \ x = u + 1$$
$$du = dx$$

$$= \int \frac{19 + 96(u+1)}{u^2 + 1} \, du$$

$$= \int \frac{19}{u^2 + 1} du + \int \frac{96u}{u^2 + 1} du + \int \frac{96}{u^2 + 1} du$$

$$= 19 \tan^{-1} u + \int \frac{96u}{u^2 + 1} du + 96 \tan^{-1} u$$

After doing the initial u substitution, we find we have 3 integrals to solve, one of which requires another u substitution. Since we've already used the variable u in this problem, we use v here.

Solve $\int \frac{96u}{u^2 + 1} du$

$$v = u^2 + 1$$

$$dv = 2u \, du$$

$$\int \frac{96u}{u^2 + 1} du = \int \frac{96u}{v}\left(\frac{dv}{2u}\right) = \int \frac{48}{v} dv = 48 \ln|v| = 48 \ln|u^2 + 1|$$

$$\therefore \quad \frac{19 + 96x}{(x-1)^2 + 1} dx = 19 \tan^{-1} u + \left(48 \ln|u^2 + 1|\right) + 96 \tan^{-1} u$$

$$= 19 \tan^{-1}(x-1) + \left(48 \ln|(x-1)^2 + 1|\right) + 96 \tan^{-1}(x-1)$$

3. $\int \frac{1}{\sqrt{x^2 + 4x + 36}} dx$

This problem is about as bad as completing the square problems get. We start with our simple transformation, begin solving, then notice that we need to do trig substitution to finish it off. Ack!

i. $x^2 + 4x + 36 = 1\left[x + \dfrac{4}{2(1)}\right]^2 + 36 - \dfrac{(4)^2}{4(1)} = (x+2)^2 + 32$

ii. $\int \frac{1}{\sqrt{(x+2)^2 + 32}} dx$

$$u = x + 2$$

$$du = dx$$

$$= \int \frac{1}{\sqrt{u^2 + 32}} du$$

Trig Substitution required.

(i) $\sqrt{u^2 + 32}$ $\therefore$ Substitute $u = \sqrt{32} \tan\theta$, $du = \sqrt{32} \sec^2 \theta \, d\theta$

$$\int \frac{1}{\sqrt{u^2 + 32}} du = \int \frac{1}{\sqrt{\left(\sqrt{32} \tan\theta\right)^2 + \left(\sqrt{32}\right)^2}} \left(\sqrt{32} \sec^2 \theta \, d\theta\right)$$

(ii)
$$= \int \frac{1}{\left(\sqrt{32}\right)\sqrt{\sec^2\theta}}\left(\sqrt{32}\,\sec^2\theta\,d\theta\right)$$

$$= \int \sec\theta\,d\theta = \ln|\sec\theta + \tan\theta| + C$$

(iii) Convert back to u: $u = \sqrt{32}\tan\theta$, $\tan\theta = \dfrac{u}{\sqrt{32}}$

$$\ln|\sec\theta + \tan\theta| + C = \ln\left|\frac{\sqrt{u^2 + 32}}{\sqrt{32}} + \frac{u}{\sqrt{32}}\right| + C = \ln\left|\frac{\sqrt{(x+2)^2 + 32}}{\sqrt{32}} + \frac{x+2}{\sqrt{32}}\right| + C$$

Additional Problems   8.4
Evaluate the following integrals.

1. $\displaystyle\int \frac{1}{\sqrt{-x^2 + 2x + 16}}\,dx$ 2. $\displaystyle\int \frac{1}{x^2 - 9x + 12}\,dx$ 3. $\displaystyle\int \frac{1}{\sqrt{2x^2 + 4x + 32}}\,dx$

8.5 Partial Fractions (and Long Division)

We now arrive at the most complicated of the seven integration techniques introduced in this chapter--*partial fractions (and long division)*. Never in the history of calculus have so many screwed up so much, as have those who have attempted to implement this gem. "Why have they flailed?" you ask. Because they never got the step-by-step rules down. Solving integrals with this technique is not a simple three-step-and-out deal like the other techniques we've learned. No, this one can get quite nasty and convoluted if you don't have a plan. Fortunately, we do.

Method for solving Partial Fractions:

1. Use this technique when you have a function of the form, $g(x)/f(x)$, and no other technique seems applicable. Usually both $g(x)$ and $f(x)$ will be a collection of polynomials and they'll be factorable at least down to powers of 2--i.e., to the forms:

$$\left(ax^2 + bx + c\right) \text{ and / or } (dx + e), \text{ where } a, b, c, d, \text{ and } e \text{ are constants.}$$

For example, the integral: $\displaystyle\int \frac{2x + 3}{(x+2)^2\left(2x^2 + 8x - 3\right)^3}\,dx$; is in the required

form for this technique, and is already factored.

2. In order to use this technique, the degree of $g(x)$ must be less than the degree of $f(x)$, so check to make sure it is. In our example this is true, so we're OK. However, if the degree of $g(x)$ is not less than $f(x)$, use *long division* to make it so ("Number One")--the examples below illustrate how this is done. Note, sometimes after doing long division you get lucky and are able to stop right there and solve. Usually, though, you'll have to proceed to Step #3.

3. Factor $f(x)$ down to products of $(dx + e)$, and/or irreducible quadratics in the form $(ax^2 + bx + c)$. In our example, the problem was given in the factored form--this is usually the case for the more complicated problems.

4. Key step: Look at the factored form of $f(x)$. Take each factor and express it as a denominator with an unknown(s) in its numerator as follows:

For all $(dx + e)^n$, write: $\dfrac{A_1}{(dx + e)} + \dfrac{A_2}{(dx + e)^2} + ... + \dfrac{A_n}{(dx + e)^n}$

For all $(ax^2 + bx + c)^m$, write: $\dfrac{B_1 x + C_1}{(ax^2 + bx + c)} + \dfrac{B_2 x + C_2}{(ax^2 + bx + c)^2} + ... + \dfrac{B_m x + C_m}{(ax^2 + bx + c)^m}$

In our above example, we thus write:

$$\frac{A_1}{(x+2)} + \frac{A_2}{(x+2)^2} + \frac{B_1 x + C_1}{(2x^2 + 8x - 3)} + \frac{B_2 x + C_2}{(2x^2 + 8x - 3)^2} + \frac{B_3 x + C_3}{(2x^2 + 8x - 3)^3}$$

5. Add all of these terms together, and then set them equal to the factored form of $g(x)/f(x)$ in the integral. In our example we would have:

$$\frac{2x + 3}{(x+2)^2 (2x^2 + 8x - 3)^3} = \frac{A_1}{(x+2)} + \frac{A_2}{(x+2)^2} + \frac{B_1 x + C_1}{(2x^2 + 8x - 3)}$$

$$+ \frac{B_2 x + C_2}{(2x^2 + 8x - 3)^2} + \frac{B_3 x + C_3}{(2x^2 + 8x - 3)^3}$$

6. Get all of the added terms together under one denominator by multiplying by the lowest common denominator. (This is where it gets ugly). In our example we would get:

$$\frac{2x + 3}{(x+2)^2 (2x^2 + 8x - 3)^3} =$$

$$\frac{A_1(x+2)(2x^2 + 8x - 3)^3 + A_2(2x^2 + 8x - 3)^3 + (B_1 x + C_1)(x+2)^2(2x^2 + 8x - 3)^2 + (B_2 x + C_2)(x+2)^2(2x^2 + 8x - 3) + (B_3 x + C_3)(x+2)^2}{(x+2)^2 (2x^2 + 8x - 3)^3}$$

7. Now, set the numerators equal to each other and solve for the unknown constants-- usually fairly simple, though in this example near impossible. Note: You're never going to get a problem to solve that's as hard as the one we're showing you here because the algebra in this step is heinous. The reason for showing you this problem is to clarify how to set these types of problems up--i.e., how to do step #4. That's the key step. Once you get by that, the rest is algebra.

8. Once you know the values for the constants--e.g., A_1, A_2, B_1, C_1, etc.--substitute them back into the equation written in Step #5.

9. Integrate both sides of the equation--since they are equal you can simply integrate the side you just created. The integration at this point is usually trivial. The examples below will clearly illustrate this.

Note: It's really important that you learn and follow these nine steps when you're trying to solve integrals using Partial Fractions. If you don't, you'll never make it.

Examples   **8.5**

Evaluate the following integrals.

1. $\int \dfrac{x^2 - 1}{x^3 + 2x^2 + x + 2} dx$

Our first partial fractions example. Note that these problems are long, but if you follow the step-by-step guidelines, they're not difficult. You get into trouble when you forget the steps.

[1] Yes, use partial fractions.

[2] $g(x) < f(x)$ in terms of magnitude

[3] $\int \dfrac{x^2 - 1}{x^3 + 2x^2 + x + 2} dx = \int \dfrac{x^2 - 1}{(x^2 + 1)(x + 2)} dx$

To review, we know that the denominator is higher magnitude than the numerator because its highest power is 3, while the highest power in the numerator is only 2.

[4] $\dfrac{Bx + C}{x^2 + 1}$, $\dfrac{A_1}{x + 2}$

[5] $\dfrac{x^2 - 1}{(x^2 + 1)(x + 2)} = \dfrac{A_1}{x + 2} + \dfrac{Bx + C}{x^2 + 1}$

[6] $\dfrac{x^2 - 1}{(x^2 + 1)(x + 2)} = \dfrac{A_1(x^2 + 1) + (Bx + C)(x + 2)}{(x + 2)(x^2 + 1)}$

[7] $x^2 - 1 = A_1 x^2 + A_1 + Bx^2 + 2Bx + Cx + 2C$

$\quad A_1 x^2 + Bx^2 = x^2 \quad \therefore \ A_1 + B = 1$

$\quad 2Bx + Cx = 0 \qquad \therefore 2B + C = 0$

$\quad A_1 + 2C = -1$

This step sometimes throws people. The setting up of the equations is usually no problem, but the solving is.

3 Equations with 3 unknowns, $\therefore$ we can solve:

$A_1 = 1 - B$

$(1 - B + 2C) = -1$

$B = 2 + 2C$

$2(2 + 2C) + C = 0$

$5C = -4$

$C = -\dfrac{4}{5}, \quad B = 2 - \dfrac{8}{5} = \dfrac{2}{5}, \ A_1 = \dfrac{3}{5}$

Note that in solving for these constants all we're doing is taking the 3 equations above, and using substitution. Just follow it through. It looks really complex, but it's just simple algebra.

[8] $\dfrac{x^2 - 1}{(x^2 + 1)(x + 2)} = \dfrac{\left(\dfrac{3}{5}\right)}{x + 2} + \dfrac{\left(\dfrac{2}{5}x - \dfrac{4}{5}\right)}{x^2 + 1}$

Once we get to this point we're home free. We plug in the constants the and integrate.

[9] $\displaystyle\int \dfrac{\left(\dfrac{3}{5}\right)}{x + 2}dx + \int \dfrac{\left(\dfrac{2}{5}x\right)}{x^2 + 1}dx + \int \dfrac{\left(-\dfrac{4}{5}\right)}{x^2 + 1}dx$

$\qquad (1) \qquad\qquad (2) \qquad\qquad (3)$

Note: In completing step #9, we have three integrals to solve, hence we label them (1), (2) and (3) to keep track of them when we solve.

We can't emphasise this enough--you must stay organized when using this techique or you'll crash and burn.

(1) $\displaystyle\int \dfrac{\left(\dfrac{3}{5}\right)}{x + 2}dx = \dfrac{3}{5}\ln|x + 2|$

(2) $\displaystyle\int \dfrac{\left(\dfrac{2}{5}x\right)}{x^2 + 1}dx$

$\qquad u = x^2 + 1$

$\qquad du = 2x\,dx$

$= \dfrac{2}{5}\displaystyle\int \dfrac{x}{u}\left(\dfrac{du}{2x}\right) = \dfrac{1}{5}\ln|u| = \dfrac{1}{5}\ln|x^2 + 1|$

(3) $\displaystyle\int \dfrac{\left(-\dfrac{4}{5}\right)}{x^2 + 1}dx = -\dfrac{4}{5}\tan^{-1}(x^2 + 1)$

$$\therefore \int \frac{x^2 - 1}{x^3 + 2x^2 + x + 2}\, dx = \frac{3}{5}\ln|x + 2| + \frac{1}{5}\ln\left|x^2 + 1\right| - \frac{4}{5}\tan^{-1}\left(x^2 + 1\right) + C$$

2. $\displaystyle \int \frac{4x^3 + 3x^2 + 2x + 1}{x^2 + 3x - 10}\, dx$

[1] Use partial fractions.

[2] $g(x) > f(x)$ $\therefore$ use long division

$$\begin{array}{r}
4x - 9 \\
\left(x^2 + 3x - 10\right)\overline{)4x^3 + 3x^2 + 2x + 1} \\
\underline{4x^3 + 12x^2 - 40x} \\
-9x^2 + 42x + 1 \\
\underline{-9x^2 - 27x + 90} \\
69x - 89
\end{array}$$

$$\therefore \int \frac{4x^3 + 3x^2 + 2x + 1}{x^2 + 3x - 10}\, dx = \int \left[4x - 9 + \frac{69x - 89}{x^2 + 3x - 10}\right] dx$$

$$= \int 4x\, dx + \int (-9)dx + \int \frac{69x - 89}{x^2 + 3x - 10}\, dx$$

$$= 2x^2 - 9x + \int \frac{69x - 89}{x^2 + 3x - 10}\, dx$$

Solve the remaining integral using partial fractions.

[3] $\displaystyle \int \frac{69x - 89}{x^2 + 3x - 10}\, dx = \int \frac{69x - 89}{(x + 5)(x - 2)}\, dx$

[4] $\dfrac{A_1}{x + 5}$, $\dfrac{A_2}{x - 2}$

[5] $\dfrac{69x - 89}{x^2 + 3x - 10} = \dfrac{A_1}{x + 5} + \dfrac{A_2}{x - 2}$

[6] $\dfrac{69x - 89}{x^2 + 3x - 10} = \dfrac{A_1(x - 2) + A_2(x + 5)}{(x + 5)(x - 2)}$

[7] $69x - 89 = A_1 x - 2A_1 + A_2 x + 5A_2$

$\therefore$ $69 = A_1 + A_2$

$-89 = -2A_1 + 5A_2$

In our second example, we see a situation that requires the use of long division. Note that long division with functions is just like what you did in arithmetic class in elementary school, just with functions. Not hard once you do one or two.

Once the long division is finished, this problem is very straight forward--no tricks. Just follow through the steps, stay organized, and don't panic.

Pay close attention to Step #4 in all the examples. If you get through Step #4, you're basically home free. The only thing standing between you and success at that point is algebra.

$$\therefore \; A_1 = 69 - A_2$$

$$-89 = -2(69 - A_2) + 5A_2$$

$$7 = A_2$$

$$\therefore \; A_1 = 62$$

[8] $\dfrac{69x - 89}{x^2 + 3x - 10} = \dfrac{62}{x+5} + \dfrac{7}{x-2}$

[9] $\displaystyle\int \dfrac{62}{x+5}dx + \int \dfrac{7}{x-2}dx = 62\ln|x+5| + 7\ln|x-2|$

$\therefore \displaystyle\int \dfrac{4x^3 + 3x^2 + 2x + 1}{x^2 + 3x - 10}dx = 2x^2 - 9x + 62\ln|x+5| + 7\ln|x-2| + C$

3. $\displaystyle\int \dfrac{x^2 + 3}{(x+1)^2(x-2)}dx$

In this problem, the biggest thing is Step #4, where we divide the fraction up to establish our unknown constants. Make sure you understand what we've done-- review Step #4 as outlined on page 177.

[1] Use partial fractions.

[2] $g(x) < f(x)$

[3] Done.

[4] $\dfrac{A_1}{x+1}, \;\; \dfrac{A_2}{(x+1)^2}, \;\; \dfrac{A_3}{x-2}$

[5] $\dfrac{x^2 + 3}{(x+1)^2(x-2)} = \dfrac{A_1}{x+1} + \dfrac{A_2}{(x+1)^2} + \dfrac{A_3}{x-2}$

[6] $\dfrac{x^2 + 3}{(x+1)^2(x-2)} = \dfrac{A_1(x+1)(x-2) + A_2(x-2) + A_3(x+1)^2}{(x+1)(x+1)^2(x-2)}$

[7] $x^2 + 3 = A_1x^2 - A_1x - 2A_1 + A_2x - 2A_2 + A_3x^2 + 2A_3x + A_3$

$\therefore \;\; x^2 = A_1x^2 + A_3x^2 \qquad\qquad \therefore \; 1 = A_1 + A_3 \qquad (1)$

$\quad\;\; 0 = -A_1x + A_2x + 2A_3x \qquad \therefore \; 0 = -A_1 + A_2 + 2A_3 \;\; (2)$

$\quad\;\; 3 = -2A_1 - 2A_2 + A_3 \qquad\qquad\qquad\qquad\qquad (3)$

(1) $A_1 = 1 - A_3$

(2) $0 = -(1 - A_3) + A_2 + 2A_3$

$0 = 3A_3 + A_2 - 1$

(3) $3 = -2(1 - A_3) - 2(1 - 3A_3) + A_3$

$A_3 = \dfrac{7}{9};$ $\quad \therefore \ A_2 = -\dfrac{4}{3};$ $\quad A_1 = \dfrac{2}{9}$

[8] $\dfrac{x^2 + 3}{(x+1)^2(x-2)} = \dfrac{\left(\dfrac{2}{9}\right)}{x+1} + \dfrac{\left(-\dfrac{4}{3}\right)}{(x+1)^2} + \dfrac{\left(\dfrac{7}{9}\right)}{x-2}$

[9] $\displaystyle\int \dfrac{\left(\dfrac{2}{9}\right)}{x+1}dx + \int \dfrac{\left(-\dfrac{4}{3}\right)}{(x+1)^2}dx + \int \dfrac{\left(\dfrac{7}{9}\right)}{x-2}dx = \dfrac{2}{9}\ln|x+1| + \dfrac{4}{3}\left(\dfrac{1}{x+1}\right) + \dfrac{7}{9}\ln|x-2| + C$

Additional Problems 8.5

Evaluate the following integrals.

1. $\displaystyle\int \dfrac{2x+6}{(x+6)(x-6)}dx$

2. $\displaystyle\int \dfrac{4x^2 + x + 1}{(x+2)^2(x-1)}dx$

3. $\displaystyle\int \dfrac{8x^3 + 2x^2 + 4}{(x^2+2)^2}dx$

4. $\displaystyle\int \dfrac{x^5}{(x^2+4)^2}dx$

8.6 Hand-Waving and Trickery

We're through the worst of it now. At this point, we're dealing with integrals that are ridiculous. You just don't see too many of these once you're out there in the real world. To solve this final collection, we employ technique number six, *hand-waving and trickery.*

The hand-waving and trickery technique is used in situations where radicals of higher order are found. For example, suppose you run into the integral:

$$\int 3x^3 \left(\sqrt[3]{x^2 + 5}\right)dx$$

If you play around with this integral for a bit, you'll see that none of the techniques we've learned so far will solve it--at least not easily. So what do we do? We do a standard substitution and massage it a little. Observe:

$u = \sqrt[3]{x^2 + 5}$

$u^3 = x^2 + 5$, and thus $x^2 = u^3 - 5$

Taking the derivative of both sides of the latter equation, $2x\, dx = 3u^2\, du$

therefore, $x\, dx = \dfrac{3}{2} u^2\, du$

Now substitute the various components back into the integral:

$$\int 3x\left(u^3 - 5\right)u\left(\frac{3}{2x}u^2\, du\right) = \int \frac{9}{2}\left(u^3 - 5\right)u^2\, du$$

and solve.

Alternatively, you could have in your original substitution used, $u = x^2 + 5$. The result would have been the same, but slightly different massaging would have gone on.

So, if you look back on this example and on to the ones to follow, you'll see that basically what this hand-waving and trickery technique is telling you to do is make a normal substitution and then algebraically manipulate the substitution equations until you get a reasonable integral. There are no set rules for solving these types of integrals. Each one will be a little different. The only thing you can really do is be aware that they exist, and practice on a few until you get comfortable with dealing with them. The examples that follow should help. Good luck.

Examples   **8.6**
Evaluate the following integrals.

1. $\displaystyle \int \frac{8x^3}{\sqrt[4]{x^2 + 8}}\, dx$

We start this problem off by picking the denominator as our substitution. Once we do this, we work the substitution equation around until we can cover everything in the integral. So how, specifically, do we cover everything in the integral? By substituting various parts of these substitution equations into the integral.

$u = \sqrt[4]{x^2 + 8}$

$u^4 = x^2 + 8 \qquad\qquad x^2 = u^4 - 8$

$4u^3\, du = 2x\, dx$

$\displaystyle \int \frac{8x^3}{\sqrt[4]{x^2 + 8}}\, dx = \int \frac{8\left(u^4 - 8\right)x}{u}\left(\frac{4u^3}{2x}\, du\right)$

$\qquad\qquad = \int 16\left(u^4 - 8\right)u^2\, du$

$\qquad\qquad = \int \left(16u^6 - 128u^2\right)du$

$\qquad\qquad = \dfrac{16}{7}u^7 - \dfrac{128}{3}u^3 = \dfrac{16}{7}\left(\sqrt[4]{x^2 + 8}\right)^7 - \dfrac{128}{3}\left(\sqrt[4]{x^2 + 8}\right)^3 + C$

2. $\int \dfrac{1}{4+\sqrt{x}}\,dx$

As with the last problem, we pick the denominator as our substitution, work those equations around, then cover everything in the integral.

$u = 4 + \sqrt{x}$

$u - 4 = \sqrt{x}$

$(u-4)^2 = x$

$2(u-4)du = dx$

$\therefore \ \int \dfrac{1}{u}[2(u-4)]du = \int \dfrac{2u-8}{u}\,du$

$\qquad\qquad = \int 2\,du - 8\int \dfrac{1}{u}\,du$

$\qquad\qquad = 2u - 8\ln|u| = 2(4+\sqrt{x}) - 8\ln\left|4+\sqrt{x}\right| + C$

3. $\int x^2\left(\sqrt[3]{x+4}\right)dx$

$u = \sqrt[3]{x+4}$

$u^3 = x+4 \qquad x = u^3 - 4 \qquad x^2 = \left(u^3-4\right)^2$

$3u^2\,du = dx$

$\int x^2\left(\sqrt[3]{x+4}\right)dx = \int \left(u^3-4\right)^2 u\left(3u^2\,du\right)$

$\qquad\qquad = \int \left(u^6 - 8u^3 + 16\right)3u^3\,du$

$\qquad\qquad = \int \left(3u^9 - 24u^6 + 48u^3\right)du$

$\qquad\qquad = \dfrac{3}{10}u^{10} - \dfrac{24}{7}u^7 + 12u^4$

$\qquad\qquad = \dfrac{3}{10}\left(\sqrt[3]{x+4}\right)^{10} - \dfrac{24}{7}\left(\sqrt[3]{x+4}\right)^7 + 12\left(\sqrt[3]{x+4}\right)^4 + C$

Additional Problems 8.6

Solve the following integrals.

1. $\int \dfrac{19x+96}{\sqrt[3]{x-1}}\,dx$

2. $\int \dfrac{5x-4}{3+\sqrt{x}}\,dx$

Chapter 9

L'Hospital's Rule & a Fourth Semester Math Advertisement

In this chapter we say good-bye to our study of integrals and begin a four chapter tour of an assortment of calculus applications. These applications will in some cases be mathematical, in others practical. The important thing to remember is that we've completed our initial study of the three mathematical operations that comprise calculus-- limits, derivatives, and integrals--and are now going to start using them. Hence, in this chapter we begin by turning our attention to a technique which utilizes derivatives to solve limits, and end with a preview of fourth semester math-i.e., differential equations.

Congratulations are in order! The fact that you're reading this sentence indicates that you have, for all intents and purposes, successfully conquered first year calculus. "How can we say this with still five chapters to go?" you ask. Simple. At this point the three mathematical operations that comprise calculus--limits, derivatives, and integrals--along with all the little nuances that go with them, have been covered. All that's left to do now is learn how to apply these operations to various mathematical and real world situations. This is not hard. Sure there will be some topics ahead that won't be all that fun, but we can assure you there's nothing lying before us that even comes close to approaching the difficulty of the material we slaved over in chapters six and eight. That was serious big time mental gymnastics. This, by comparison, is a walk in the park. If you've survived OK thus far, you've got nothing to worry about. Just remember, what we'll be covering in the next four chapters is just applications of the three mathematical operations we spent the last seven chapters studying. Don't make it harder than it is. Just sit back, relax, and enjoy.

9.1 L'Hospital's Rule

We all know how to solve limits, right? The first thing we do is plug the specified number into the function and see what happens. As you will recall, if we get a number, that is the limit of the function, if we get a number divided by zero, the limit does not exist, and if we get zero divided by zero we must either factor, reduce in a different manner, or graph in order to solve. In the case of infinite limits, the same applies except we have the additional complication of getting infinity divided by infinity and having to reduce that.

Clearly, of the scenarios outlined above, the majority of the problem solving headaches come when we get zero divided by zero (or infinity divided by infinity) and have to use various techniques to simplify the functions being operated on. Up until now, we've been using mainly brute force to do this. Fortunately, there is an easier way, L'Hospital's rule. Here it is:

L'Hospital's Rule:

IF you are given two functions, $f(x)$ and $g(x)$, and you desire to take the limit,

$$\lim_{x \to a} \frac{g(x)}{f(x)}$$

AND you find that you get either zero divided by zero, or infinity divided by infinity, upon plugging in a, THEN you can use L'Hospital's rule, which says:

$$\lim_{x \to a} \frac{g(x)}{f(x)} = \lim_{x \to a} \frac{g'(x)}{f'(x)}$$

Translation: When you're given a limit that's in the form $g(x)/f(x)$ that yields either zero divided by zero or infinity divided by infinity upon plugging in the specified number, a, you can take the derivative of each function--separately--and then try to solve again. If you still get zero divided by zero or infinity divided by infinity, you can repeat the process.

IMPORTANT: If you take the derivative of one of the functions, you must take the derivative of the other, and you can only do this when plugging in the desired input yields either zero divided by zero or infinity divided by infinity.

Examples 9.1

Find the limits in the following problems.

1. $\lim_{x \to -2} \dfrac{x+2}{x^3 + 8}$

$\lim_{x \to -2} \dfrac{x+2}{x^3 + 8} = \lim_{x \to -2} \dfrac{1}{3x^2} = \dfrac{1}{12}$

We look at this problem, mentally plug the -2 into the function, note that we would in fact get zero divided by zero, and so use L'Hospital's rule. In doing so, we take the derivative of the function in the numerator, which is equal to 1, then the derivative of the function in the denominator, which is equal to 3 x^2, and then solve. Easy!

2. $\lim\limits_{x \to -2} \dfrac{4x^3 + 32}{4x^4 - 64}$

Plugging in the specified number yields 0/0, so we use L'Hospital's rule: take the derivatives, simplify the resulting expression, then try to solve again. This time out pops an easy answer.

$$\lim_{x \to -2} \frac{4x^3 + 32}{4x^4 - 64} = \lim_{x \to -2} \frac{12x^2}{16x^3} = \lim_{x \to -2} \frac{3}{4x} = -\frac{3}{8}$$

3. $\lim\limits_{x \to 0} \dfrac{\dfrac{1}{\sqrt{4-x}} - \dfrac{1}{2}}{x}$

Slightly more complex derivative for the function in the numerator, but still infinitely easier than using the conjugate method, yes? If you don't believe us, see Example #16 on page 21.

$$\lim_{x \to 0} \frac{\dfrac{1}{\sqrt{4-x}} - \dfrac{1}{2}}{x} = \lim_{x \to 0} \frac{\left(\dfrac{\left(\sqrt{4-x}\right)(0) - (1)\left(\dfrac{1}{2}\right)(4-x)^{-\frac{1}{2}}(-1)}{\left(\sqrt{4-x}\right)^2} - 0 \right)}{1}$$

$$= \lim_{x \to 0} \frac{1}{2\left(\sqrt{4-x}\right)(4-x)} = \frac{1}{2(2)(4)} = \frac{1}{16}$$

Additional Problems   9.1
Find the limits in the following problems.

1. $\lim\limits_{x \to 2} \dfrac{3x^2 - 10x + 8}{x^3 - 2x^2 - 2x + 4}$

4. $\lim\limits_{x \to 0} \dfrac{4 - \sqrt{16 + x}}{x}$

7. $\lim\limits_{x \to 0} \dfrac{\sin x - x}{\tan x - x}$

2. $\lim\limits_{x \to 1} \dfrac{x^3 - 1}{\sqrt{x} - 1}$

5. $\lim\limits_{x \to 1} \dfrac{-2 + \sqrt{x + 3}}{x}$

8. $\lim\limits_{x \to 0} \dfrac{x - \sin x}{3x^2}$

3. $\lim\limits_{x \to 1} \dfrac{x^3 - 1}{x^2 - 1}$

6. $\lim\limits_{x \to \frac{\pi}{2}} \dfrac{\sin x + 4x - 12}{\left(x - \dfrac{\pi}{2} \right)}$

9. $\lim\limits_{x \to \infty} \dfrac{4x \ln x}{x + \ln x}$

9.2 Trickery in the L'Hospital Rule

It's important to remember that one of the major constraints of L'Hospital's rule is that you have one function divided by another--other combinations are not acceptable. However, *there's no saying you can't perform a little trickery of your own to get two functions into this form.* For example, suppose you see two functions multiplied together as in $f(x)g(x)$. At first glance you might think L'Hospital's rule won't work on them, but if do the following, you can in fact get them in the required divided form:

$$f(x) \cdot g(x) = \frac{f(x)}{\left(\dfrac{1}{g(x)}\right)} = \frac{g(x)}{\left(\dfrac{1}{f(x)}\right)}$$

Similarly, if you have one function raised to the power of another function, you can do the following in order to solve via L'Hospital's rule:

Given a function in the form: $y = g(x)^{f(x)}$, do the following:

[1] $\ln y = \ln g(x)^{f(x)}$

[2] $\ln y = f(x) \ln g(x)$

[3] $\ln y = \dfrac{\ln g(x)}{\left(\dfrac{1}{f(x)}\right)}$

[4] Take the limit of this new expression found in step #3 using L'Hospital's rule.

[5] Convert the answer obtained in step #4 - -i.e., $e^{\text{answer in step #4}}$. See Example #3 below.

NOTE: We can't emphasize this enough. Besides having to have one function divided by the other, in order to use L'Hospital's rule you must get zero divided by zero or infinity divided by infinity to use the rule. Don't forget this requirement. The form is not enough.

Examples   9.2
Find the limits if they exist.

1. $\lim\limits_{x \to 0^+} x \ln x$

$\lim\limits_{x \to 0^+} x \ln x = (0)(\infty) \;\; \therefore \;\; \text{indeterminant}$

$\lim\limits_{x \to 0^+} x \ln x = \lim\limits_{x \to 0^+} \dfrac{\ln x}{\left(\dfrac{1}{x}\right)} = \lim\limits_{x \to 0^+} \dfrac{\left(\dfrac{1}{x}\right)}{\left(-\dfrac{1}{x^2}\right)} = \lim\limits_{x \to 0^+} (-x) = 0$

We try to solve this problem by first plugging in the specified number. When we do this, however, we get an indeterminate answer--zero multiplied by infinity. Hence, we use the first algebraic manipulation outlined above to enable us to solve this problem via L'Hospital's rule.

2. $\lim\limits_{x \to 0} x \csc x$

$\lim\limits_{x \to 0} x \csc x = (0)(\infty) \quad \therefore \text{ indeterminant}$

$\lim\limits_{x \to 0} x \csc x = \lim\limits_{x \to 0} \dfrac{x}{\left(\dfrac{1}{\csc x}\right)}$

$= \lim\limits_{x \to 0} \dfrac{x}{\sin x} = \lim\limits_{x \to 0} \dfrac{1}{\cos x} = \dfrac{1}{1} = 1$

Same deal as the last example. We plug in the specified number, get an indeterminate form, then modify and solve via L'Hospital's rule.

After doing the necessary manipulation of the functions, then taking the derivatives, we come up with an answer of 1. Surprise!

3. $\lim\limits_{x \to 0} (1 + 2x)^{\frac{3}{x}}$

$\lim\limits_{x \to 0} (1 + 2x)^{\frac{3}{x}} = 1^{\infty} \quad \therefore \text{ indeterminant}$

$\therefore \quad \text{say } y = (1 + 2x)^{\frac{3}{x}}$

$\ln y = \ln(1 + 2x)^{\frac{3}{x}}$

$\ln y = \left(\dfrac{3}{x}\right)\ln(1 + 2x)$

$\ln y = \dfrac{\ln(1 + 2x)}{\left(\dfrac{x}{3}\right)} = \dfrac{3\ln(1 + 2x)}{x}$

Here's an example of one of the types of problems that requires major manipulating. We showed you the steps for solving it above on page 189. It's not too bad as long as you remember your rules for exponents. Even if you don't, though, you can get by OK by just memorizing the patterns you'll find in this example.

$\therefore \quad \lim\limits_{x \to 0} \ln y = \lim\limits_{x \to 0} \dfrac{3\ln(1 + 2x)}{x} = \dfrac{3\ln(1 + 0)}{0} = \dfrac{3(0)}{0} = \dfrac{0}{0}$

$\therefore \quad \lim\limits_{x \to 0} \dfrac{3\ln(1 + 2x)}{x} = \dfrac{3\left(\dfrac{1}{1 + 2x}(2)\right)}{1} = \lim\limits_{x \to 0} \dfrac{6}{1 + 2x} = \dfrac{6}{1} = 6$

$\therefore \quad \lim\limits_{x \to 0} \ln y = 6$

$\therefore \quad \lim\limits_{x \to 0} y = e^6$

$\therefore \quad \lim\limits_{x \to 0} (2 + 2x)^{\frac{3}{x}} = e^6$

Here's where the big conversion is--the key to the problem. We finish solving the limit of the modified form, but we need to invert the ln result to get our answer. Look back to Chapter 1 if you need to review the rules for doing this.

Additional Problems   9.2

Find the limits if they exist.

1. $\lim\limits_{x \to 0^+} 5x^4 \ln x$

2. $\lim\limits_{x \to 0} 9x^2 \cot x$

3. $\lim\limits_{x \to \infty} \dfrac{3x^2 - 4}{x^3 + 5x^2 - 1}$

4. $\lim\limits_{x \to 0^+} \left(e^x - 1\right)^{x^2}$

9.3 Exponential Growth and Decay

It's time for a preview of fourth semester--i.e., a look a *differential equations*. Obvious question: What are differential equations? *Differential equations are simply equations with derivatives in them.* Hence we find that fourth semester math is firmly rooted in first and second semester math. Everything builds. So what do differential equations have to do with this section heading, "Exponential Growth and Decay"? Well, both phenomena, exponential growth and exponential decay, can be modeled with differential equations. Hence, exponential growth and exponential decay are nice examples which we can use to introduce this topic. We begin our discussion then, with the following example:

Suppose an investment banker tells us that she's put together an investment portfolio that will increase investors' money at a rate proportional to the amount they put in. In order to calculate how much a certain amount of money will grow over a given period of time, we decide to derive an equation to describe her portfolio as follows:

The first thing we do is examine the given information: Our invested money grows at a rate proportional to the amount present, so:

1. Let y = value of the investment

2. $\therefore$ $\dfrac{dy}{dt}$ = the rate at which the investment grows

3. We know $\dfrac{dy}{dt}$ is proportional to y, therefore we can write:

$$\frac{dy}{dt} = hy, \text{ where } h \text{ is a proportionality constant}$$

We now have a differential equation--an equation with a derivative in it. Which is great except it's useless at this point. Why? Because we don't know anything about h or dy/dt. How do we figure out what h and dy/dt are? We solve the above equation as follows:

$$\frac{dy}{dt} = hy$$

$$\frac{dy}{y} = h\,dt$$

$$\int \frac{1}{y}\,dy = \int h\,dt$$

$$\ln|y| = ht + C$$

$$e^{\ln|y|} = e^{ht+C}$$

$$y = e^C e^{ht}$$

Don't panic. This is a real life example of a real life differential equation being put into a useful form. We're only showing this to you to give you a taste for what lies ahead in fourth semester. Relax!

Let y_o = the initial value of the portfolio

Then, $y_o = e^C e^{h(0)} = e^C$

$$\therefore \ y = y_o e^{ht}$$

We now have a useful equation. Why? Because we can find out from the investment manager, based on her past experience with this portfolio, how much a certain amount of money (y_o--the initial investment) grew (y--the final amount the investment was worth) over a given period of time (t--the time period over which the money was invested). Using this information, we can solve for the constant h. This h will be the same for this portfolio always--but ONLY for this particular portfolio. Once we have h then, we plug it into our equation and leave it there. Our equation now serves its purpose: It tells us how much a certain investment, y_o, will grow over a given period of time, t. Hence, in order to use the equation, all we have to do is plug in y_o and t and we'll get y.

Note that the differential equation we just derived is an exponential equation, hence the name exponential growth. The only difference between exponential growth and exponential decay is the value of h. If h is positive then the equation is for exponential growth, and if h is negative the equation is for exponential decay.

h, as we just alluded to, is found through a set of given conditions--a specific set of values for y, y_o, and t. In our example we continually referred to the equation as an exponential growth equation even though we didn't know the value of h. In the problems we'll be facing, usually all we'll have to do is plug various values into the equation to solve for h. Occasionally, though, we'll have to go through a derivation similar to the one above-- though not as complex--that will alter the form of the equation slightly because of the conditions given in the problem. So, make sure you understand what we did in the above derivation so you can repeat it, or slight variations of it, if need be.

Examples 9.3

1. You and your best friend are lab partners in biology. As it turns out, the class is studying bacteria, so you're all growing some in the lab. Your first assignment is to figure out how long it will take the bacteria in your culture to increase by a factor of 15 if the growth rate is exponential in nature. You note that after watching your bacteria culture for one hour their number has tripled.

We know from our derivation in the text that, $y = y_o e^{ht}$

Let y = # of bacteria in the culture

Let y_o = # of bacteria initially in the culture

at $t = 1$, $y = 3y_o$

$$\therefore \; 3y_o = y_o e^{h(1)}$$
$$3 = e^h$$
$$\ln 3 = \ln e^h$$
$$\ln 3 = h$$
$$\therefore \; y = y_o e^{t \ln 3}$$

After we get the given information translated into variable terms, we begin by solving for h by making use of the information that the bacteria in the culture tripled after 1 hour.

So, when $y = 15y_o$

$$15 = e^{t \ln 3}$$
$$15 = e^{\ln 3^t}$$
$$15 = \ln 3^t$$
$$15 = t \ln 3$$
$$\frac{15}{\ln 3} = t$$
$$t = 2.46 \text{ hours}$$

Once we have determined h, we simply start over with the equation and solve the problem using the value for h. Remember, here we're interested in how long it takes for the bacteria to increase by a factor of 15. Hence we're looking for t, y is expressed in terms of y_o, so the y_o's cancel immediately and we don't need to know there value to solve.

2. After finishing the assignment you both head off to the local hangout. As you're sitting back, chatting over a nice cold one, a news flash comes across the TV indicating two terrorists holding 1 gram of plutonium 239 are threatening to drop it somewhere in the city if their demands are not met. Pictures of the two terrorists are shown, and of course the announcement to call authorities if anyone spots them.

Knowing that plutonium 239 has a half-life of 24,360 years, you decide just for the depressing quality to calculate how long it would take for this plutonium to decay to what you estimate to be a save level of 1 microgram.

Let y = the amount of plutonium

Let y_o = the initial amount of plutonium = 1 g

$$y = y_o e^{ht}$$

$$0.5 = 1e^{h(24,360)}$$

$$\ln 0.5 = \ln e^{h(24,360)}$$

$$\ln 0.5 = 24,360h$$

$$\frac{\ln 0.5}{24,360} = h$$

$$\therefore \ h = -2.85 \times 10^{-6}$$

We begin this problem by solving for h. How do we do this. Well, we know the half life is 24,360 years--that's t. Since half-life means the time it takes to decay to half of some starting value, which in this case is 1 gram--y_o, we also know y, it's 0.5 grams.

$$y = y_o e^{ht}$$

$$1 \times 10^{-6} = 1e^{t\left(-2.85 \times 10^{-6}\right)}$$

$$\ln\left(1 \times 10^{-6}\right) = \ln e^{t\left(-2.85 \times 10^{-6}\right)}$$

$$\ln\left(1 \times 10^{-6}\right) = t\left(-2.85 \times 10^{-6}\right)$$

$$\frac{\ln\left(1 \times 10^{-6}\right)}{-2.85 \times 10^{-6}} = t$$

$$\therefore \ t = 4.86 \times 10^5 \text{ years}$$

Once we have h, we have all we need to solve for the time it will take the substance to decay to one microgram-- $1x10^{-6}$ grams (y). Remember, y_o is still 1 gram here.

Additional Problems   9.3
Solve the following story problems.

1. Needless to say, the news has put a damper on your little get together. Not feeling much like relaxing anymore, you decide to head back to your apartment. As you're getting ready to leave, however, you spot two shady looking characters over in the corner. Upon closer inspection you panic--it's the terrorists! Unfortunately before you can run over to the pay phone to call for help, the terrorists realize they've been had, stand up on a table and threaten to drop the plutonium right on the spot. Everyone freezes as the two defiantly hold up the bottle.

At this point you're pretty flipped out. Not knowing what to do, you start to feel your body shake--death was not on your list of things to experience today. Then, just as you're really getting ready to completely lose it, you notice that on the bottle they're holding are the letters Po. Thinking back to your freshman chemistry class, you realize that these terrorists have completely flailed--Po is the chemical abbreviation for polonium, probably polonium 212, which has a half-life of only 0.3 microseconds! Obviously not plutonium, whose symbol is Pu. Feeling much more comfortable, you calculate how long it would take for a gram of this stuff to reduce down to 1 microgram.

2. With the scare over and the terrorists arrested, you decide to go back for a few more cold ones. As you're sitting there, you hear some "expert" on the radio talking about the coming population problem in the world. According to this gentleman, the world's population increases at 3 percent each year. He says there's 6 billion people in the world today and that he figures the planet can only sustain 75 billion. Assuming exponential growth, if this gentleman is correct, how long before we're doomed?

3. Well that's a relief. Even if on the off chance this guy is right you figure you're safe. Anyway, before you leave you think back to your experience with the terrorists. For some reason your mind wanders on to the topic of radioactivity. You wonder what the half-life of a substance is if 10% of it decays a year.

4. In the process of letting your mind wander like that, you notice that the hot cup of coffee you were consuming in an effort to right yourself isn't quite so hot anymore. In fact, you figure that since your last sip 15 minutes ago, the stuff has cooled from 160 to 100 degrees Fahrenheit. If the temperature of the surroundings is 70 degrees, how long will it take before the coffee cools down another 20 degrees if the rate at which this coffee's cooling is directly proportional to the difference in temperature between it and the surroundings?

Note: Question #4 is a little tricky. You have to go through a new derivation to solve it.

Sequences & Infinite Series

We're about to learn how to approximate functions. This is huge. In fact, it's probably the biggest thing we learn in all of first year calculus. Why? Because approximating functions is one of the major keys to applying everything we've learned in this course to the real world. Simply put: This is where theory meets reality.

Huge! Big Time! Yes folks, in this chapter we firmly cross over the bounds of the theoretical and say hello to the real world. We are there. In this chapter we learn how to use one of the major tools that lets scientists, economists, business people, and/or any interested persons apply advanced math to the real world. Unfortunately, you probably won't realize the full significance of this chapter until you get into a laboratory or other real world setting and have a computer to play with and some real data to analyze. Nevertheless, we'll do our best to breathe some real world fresh air into this chapter to give you a taste of what lies ahead.

One note of caution before we begin. We do have some background material to slosh through before we get into the hard-core applications. Don't fret. Just hang in there. The payoff will be worth it.

10.1 Sequences--The Basic Idea

Think back for a moment to the functions we've been dealing with so far in this first year of calculus. What's the one thing they've all more or less had in common? Their domains, right? Virtually all of the functions we've dealt with have had the set of real numbers as their domain. Translation: We could essentially plug any number we wanted to into the function and get that number's corresponding output.

Well, in this section we're going to take a look at a particular group of functions whose domain is the set of *positive integers*--not the set of real numbers. Why are we doing this? Because in many real world situations the functions of interest are those with the set of positive integers as their domain. Hence, our ability to analyze these types of functions is most useful. Here's a Real World Application to show you what we mean:

Real World Application: Any time you collect data that is time dependent you always collect the data at a fixed rate--in other words at a certain frequency. Hence, functions we construct to describe the phenomena underlying time dependent data are sequences. For example, let's say we're collecting data at 60 Hz--in other words, 60 times per second. Bits of data are collected at:

1/60 sec., 2/60 sec., 3/60 sec., 4/60 sec., ...

If we form these moments in time into a set of numbers--e.g., {1/60, 2/60, 3/60, ...}-and then multiply them by the value of the frequency, in this case 60, we get a set of positive integers: {1, 2, 3, 4, ...}. Key point: If we consider this new set of numbers to be an input and the collected data they correspond to to be an output, a function can be found to relate the two groups of numbers--and this function is a sequence because its domain is a set of positive integers.

Now that you have an inkling of the usefulness--or at least where we find--functions with their domains being the set of positive integers, we need to learn some new terminology so that we can proceed with our analysis of these functions. We begin with some simple definitions:

A sequence is nothing more than a special name given to any function whose domain is the set of positive integers.

Quick Review: In Chapter 1 we defined a function as a mathematical rule that tells us how two groups of numbers are related to each other. The function accomplishes this task by pointing out that if you take one of the groups of numbers and perform some series of mathematical operations on them, you will obtain the other group. By convention, the group of numbers that gets worked on is referred to as the *domain*, the group that gets produced is referred to as the *range*, and the series of mathematical operations is the mathematical rule--i.e., the function.

So here we are concerned with functions whose domain is the set of positive integers: domain = {1, 2, 3, 4, 5, ...}. For convenience we'll use the letter n to represent elements of the domain, thus all of these functions will be expressed as $f(n)$. Carrying on with our formal notation, the ranges of these functions are written as: range = { $f(1), f(2), f(3), f(4), f(5), ...$ }.

As you can see, when writing out the ranges of these functions using this formal notation, it gets kind of cumbersome. Hence we introduce *subscript notation*, where $a_n = f(n)$ and the range is written as: range = { $a_1, a_2, a_3, a_4, a_5, ...$ }. Subscript notation results in a slight change in terminology too. Though $a_n = f(n)$, and $f(n)$ is called the function, we refer to a_n as the *general term*. Also, when we write the ranges out in subscript notation, we usually include a_n so that people can instantly see what the function is (just by looking at the range). Example: Using subscript notation we write the range as, $\{\, a_1, a_2, a_3, a_4, a_5,$ ... a_n ...$\}$, which means that if the function, $f(n)$ equals $(2n+2)$, the range is equal to, $\{\, 4, 6, 8, 10, 12, ..., 2n+2, ... \}$. Not too difficult.

So what types of problems are we going to encounter down the road with sequences? Well, there are two main types:

1. Often we'll be given a_n and asked to find certain terms in the sequence. This is trivial--the same as asking you to plug numbers into a function.

2. The other type of problem commonly posed in relation to sequences is the type that requires you to find the limit of the sequences as n approaches infinity. Since a_n is really just another way of representing the function, these problems are also trivial at this point since we've already solved reams of them in Chapter 4--the only difference here is the notation. Don't let it fool you.

Examples   10.1

Find the first 3 terms of the infinite sequences given their a_n's, and find their limit as n approaches infinity.

1. $a_n = n + 2$

To find the first three terms of the function we plug in the first three numbers in the domain and record the output-- just like a regular function.. Since the domain of these functions is the set of positive integers, the first three numbers in the domain are 1, 2, and 3 respectively. Plugging them into the function, we get their output.

$a_1 = 1 + 2 = 3$

$a_2 = 2 + 2 = 4$

$a_3 = 3 + 2 = 5$

Solving for the limit of this function is done just like in the past. No tricks, just different notation.

$$\lim_{n \to \infty} (n+2) = \infty \quad \therefore \text{ diverges}$$

2. $a_n = 17$

Don't be fooled by simple problems. Here the function is a constant--no independent variable--so we treat it here just like we did in the past.

$a_1 = 17 = 17$

$a_2 = 17 = 17$

$a_3 = 17 = 17$

$$\lim_{n \to \infty} (17) = 17 \quad \therefore \text{ converges}$$

3. $a_n = \dfrac{n}{n^2 + 1}$

Same thing as the others--just plug in 1, 2, and 3 to get the first three terms in the range.

$a_1 = \dfrac{1}{(1)^2 + 1} = \dfrac{1}{2}$

$a_2 = \dfrac{2}{(2)^2 + 1} = \dfrac{2}{5}$

$a_3 = \dfrac{3}{(3)^2 + 1} = \dfrac{3}{10}$

In solving this limit we make use of L'Hopital's rule. Much easier this way, yes?

$\lim\limits_{n \to \infty} \left(\dfrac{n}{n^2 + 1} \right) = \dfrac{1}{2n} = 0 \quad \therefore \ \text{converges}$

Additional Problems   **10.1**

Find the first 3 terms of the infinite sequences given their a_n's, and find their limit as n approaches infinity.

1. $a_n = 8n$

2. $a_n = \dfrac{n+5}{n^2 + n - 1}$

3. $a_n = \left(2 + \dfrac{1}{n} \right)^n$

4. $a_n = \left(\dfrac{2n^2}{3n - 1} - \dfrac{2n^2}{3n + 1} \right)$

5. $a_n = \dfrac{3n^5 + 1}{15e^n}$

10.2 Infinite Series--The Basic Idea

Suppose we have an infinite sequence, a_n, and we decide that we would like, as part of our analysis of it, to add all of the terms in the sequence together. Doing this gives us the following expression:

$$(a_1 + a_2 + a_3 + a_4 + a_5 + ... + a_n + ...) \qquad \text{or equivalently,} \qquad \sum_{n=1}^{\infty} a_n$$

We call this expression an *infinite series*. Infinite series are loaded with useful information. Some of the more important information that we get from them are their various sums--i.e., their *partial sums*, S_n. A partial sum is defined as follows:

$$S_n = a_1 + a_2 + a_3 + a_4 + a_5 + ... + a_n$$

$$\therefore \ S_1 = a_1$$

$$S_2 = a_1 + a_2$$

$$S_3 = a_1 + a_2 + a_3$$

$$S_4 = a_1 + a_2 + a_3 + a_4$$

etc...

Note: We can list these partial sums as a sequence of partial sums, as in $\{ S_1, S_2, S_3, ..., S_n \}$.

Real World Application: Why do we care about partial sums of infinite series? Because we can use this information to do *numerical* integration of a sequence on a computer. That's right. Using this idea of partial sums, we can take real world data that's in the form of a sequence (see previous *Real World Application* for how this is done), form it into a series and then integrate it numerically. Pretty slick. You'll learn about these *numerical methods* in your more advanced computer classes.

Another piece of information that we can get out of an infinite series is its total sum--in other words, the sum that would result if you added every term in the infinite series together. This sounds a little hokey, doesn't it. Afterall, how are we suppose to add an infinite number of numbers? Well, in actuality, we can't do it, but we can approximate it using limits in some cases. All we do is take the limit of S_n as n approaches infinity. If we get an answer, that is the total sum of the infinite series and the series is said to converge. If the limit does not exist, the total sum is infinite and the series is said to diverge.

Now, having said all this, we should point out to you that while this total sum information is theoretically useful, it's a rare day indeed when you're given an expression for S_n. Usually we're just given a_n, or we can at least find it pretty easy. S_n is a different story, and since there are other ways to determine if an infinite series converges, we don't spend any time trying to pursue expressions for S_n. It's not practical, but it is important that you be familiar with the concept.

10.3 Common Types of Series

There are five common types of series which will be beneficial for you to be aware of:

1. Geometric
2. Harmonic
3. Alternating
4. Telescoping
5. p-series

In this section we define each of these series mathematically, and give you information about when they converge and diverge. Though it won't be particularly fun, it would be

best if you memorized these definitions and bits of information thoroughly as we'll be
seeing them throughout the next two sections.

1. Geometric Series

Example: $\displaystyle\sum_{n=1}^{\infty} vr^{n-1}$; where v and r are both constants.

A geometric series converges when $|r| < 1$ and diverges when $|r| > 1$.

If a geometric series converges, then its total sum is: $\dfrac{v}{1-r}$.

2. Harmonic Series

Example: $\displaystyle\sum_{n=1}^{\infty} \dfrac{v}{b+n}$; where v and b are both constants.

All harmonic series diverge and thus have no total sum.

Also, note that b can equal zero. Thus, you'll often see harmonic series like, $\dfrac{1}{n}$

3. Alternating Series

Example: $\displaystyle\sum_{n=1}^{\infty} (-1)^{n-k} a_n$; where k is a constant and a_n is the general term (see pg 198).

An alternating series converges if the following two conditions are met:

1. $\displaystyle\lim_{n\to\infty} a_n = 0$

2. $a_{n+1} \le a_n$; for all n.

The total sum of a converging alternating series can, at best, be approximated.
This is usually done with relatively good accuracy by adding the first four to
six terms of the series.

4. Telescoping Series

Example: $\displaystyle\sum_{n=1}^{\infty} \dfrac{v}{b+2}(n+b+1)$; where v and b are both constants.

Telescoping series always converge, and their total sum is, $\dfrac{v}{b+1}$

5. p-Series

Example: $\displaystyle\sum_{n=1}^{\infty} \dfrac{1}{n^p}$; where p is a constant.

p - series converge if $p > 1$, and diverge if $p \le 1$. Sums are usually calculated
via approximation, but are rarely asked for.

Examples   **10.3**

Determine if the given infinite series converges or diverges. If it converges, find its total sum if approximation is not required.

1. $\displaystyle\sum_{n=1}^{\infty} 2\left(\frac{1}{5}\right)^{n-1}$

Pretty straight forward. The name of the game is to figure out what type of series we have, then simply use the formulas and information provided above to solve.

This is a geometric series, with $v = 2$, $r = \dfrac{1}{5}$. Therefore,

since $\left|\dfrac{1}{5}\right| < 1$, the series converges.

$$\text{Total sum} = \frac{v}{1-r} = \frac{2}{1-\left(\dfrac{1}{5}\right)} = \frac{2}{\left(\dfrac{4}{5}\right)} = \frac{5}{2}$$

2. $\displaystyle\sum_{n=1}^{\infty}\left(\frac{e}{2}\right)^{n-1}$

Couple of things. This is a geometric series--albeit somewhat disguised with v equaling 1.

This is a geometric series, with $v = 1$, $r = \dfrac{e}{2}$. Therefore,

Remember, series that diverge have no total sum.

since $\left|\dfrac{e}{2}\right| > 1$, the series diverges.

3. $\displaystyle\sum_{n=1}^{\infty}\left(\frac{e}{10}\right)^{n-1}$

Yet another geometric series. Since you'll see geometric series more than any other type, we want to make sure you're real familiar with them.

This is a geometric series, with $v = 1$, $r = \dfrac{e}{10}$. Therefore,

since $\left|\dfrac{e}{10}\right| < 1$, the series converges.

$$\text{Total sum} = \frac{v}{1-r} = \frac{1}{1-\left(\dfrac{e}{10}\right)} = \frac{10}{10-e}$$

4. $\displaystyle\sum_{n=1}^{\infty}\frac{19}{3+n}$

Whenever you see a harmonic series, rejoice! It's a well deserved break, since all harmonic series, by definition, diverge.

This is a harmonic series, $\therefore$ it diverges.

5. $\displaystyle\sum_{n=1}^{\infty}(-1)^{n-5}\left(\frac{4n^2}{3n^3-1}\right)$

For alternating series we have a lot of work to do. First, to see if the series converges, we take the limit as shown to the left-- which is no problem.

It's an alternating series, so we have 2 conditions to test:

(1) $\displaystyle\lim_{n\to\infty}\left(\frac{4n^2}{3n^3-1}\right)=\lim_{n\to\infty}\left(\frac{8n}{9n^2}\right)=\lim_{n\to\infty}\left(\frac{8}{9n}\right)=0\ \therefore\ \text{OK}$

(2) $a_{n+1}\leq a_n$

If the limit test turns out OK, then we have to test condition #2, which is a bit more difficult. Essentially, what you do after creating the inequality is plug in various numbers from the domain---usually low ones and high ones like 1, 2, 10,000, etc., to see if the expression is true or false. If it is true, then the series converges.

$$\frac{4(n+1)^2}{3(n+1)^3-1}\leq\frac{4n^2}{3n^3-1}$$

$n=1;\quad\dfrac{4(1+1)^2}{3(1+1)^3-1}\leq\dfrac{4(1)^2}{3(1)^3-1}$

$\qquad\qquad 0.6956\leq 2\quad\therefore\ \text{true}$

Here, we simply plug in two numbers--which is usually enough--an conclude that the series converges.

$n=1000;\quad\dfrac{4(1000+1)^2}{3(1000+1)^3-1}\leq\dfrac{4(1000)^2}{3(1000)^3-1}$

$\qquad\qquad 0.001332\leq 0.0013333\quad\therefore\ \text{true}$

$\therefore\ $ the series converges.

Can only approximate sum, but we'll do it anyway.

$a_1=(-1)^{1-5}\left(\dfrac{4(1)^2}{3(1)^3-1}\right)=+2$

Remember, there are no hard and fast rules about getting the sum of an alternating series, but you can approximate it via adding the first few terms together.

$a_2=(-1)^{2-5}\left(\dfrac{4(2)^2}{3(2)^3-1}\right)=-\dfrac{16}{23}$

$a_3=(-1)^{3-5}\left(\dfrac{4(3)^2}{3(3)^3-1}\right)=+\dfrac{36}{80}$

$a_4=(-1)^{4-5}\left(\dfrac{4(4)^2}{3(4)^3-1}\right)=-\dfrac{64}{191}$

$\therefore\ \text{sum}=2+\left(-\dfrac{16}{23}\right)+\dfrac{36}{80}+\left(-\dfrac{64}{191}\right)=1.419$

6. $\displaystyle\sum_{n=1}^{\infty} \frac{29}{(18+n)(n+19)}$

Rewriting, we get:

$$\sum_{n=1}^{\infty} \frac{29}{(18+n)(n+18+1)}$$

∴ We have a telescoping series,

∴ converges. Total sum $= \dfrac{v}{b+1} = \dfrac{29}{18+1} = \dfrac{29}{19}$

Pretty easy problem once you notice the rewriting trick. In the case of telescoping series, we often have to rewrite the function to see that it is, in fact, a telescoping series. Be aware.

7. $\displaystyle\sum_{n=1}^{\infty} \frac{2}{n^2 + 11n + 30}$

Rewriting, we get:

$$\sum_{n=1}^{\infty} \frac{29}{(n+5)(n+6)} = \sum_{n=1}^{\infty} \frac{29}{(n+5)(n+5+1)}$$

∴ We have a telescoping series,

∴ converges. Total sum $= \dfrac{v}{b+1} = \dfrac{2}{5+1} = \dfrac{1}{3}$

Another example of some trickery we see when dealing with telescoping series. Just some minor mental gymnastics.

8. $\displaystyle\sum_{n+1}^{\infty} \frac{1}{n^8}$

p - series, $8 > 1$, ∴ converges

p-series are very easy. All we have to do is recognize them, then use the simple test for convergence outlined above--i.e., if the exponent is greater than 1 it converges.

Don't worry about trying to calculate the total sum--it's just never done.

9. $\displaystyle\sum_{n+1}^{\infty} \left(\frac{1}{n^{\frac{1}{2}}} \right)$

p - series, $\dfrac{1}{2} < 1$, ∴ diverges

Additional Problems **10.3**

Determine if the following series converge or diverge. It the series converges, find its sum if approximation is not required.

1. $\displaystyle\sum_{n=1}^{\infty} \frac{21}{(2)^{n-1}}$

2. $\displaystyle\sum_{n=1}^{\infty} \left(-\frac{1}{3}\right)^{n-1}$

3. $\displaystyle\sum_{n=1}^{\infty} (5)^{n-1}(19)^{-n}$

4. $\displaystyle\sum_{n=1}^{\infty} 5(3)^{n-1}$

5. $\displaystyle\sum_{n=1}^{\infty} \frac{1}{1+n}$

6. $\displaystyle\sum_{n=1}^{\infty} \frac{-19}{36+n}$

7. $\displaystyle\sum_{n=1}^{\infty} \frac{3}{n}$

8. $\displaystyle\sum_{n=1}^{\infty} (-1)^{n-1}\left(\frac{5}{15n^2+1}\right)$

9. $\displaystyle\sum_{n=1}^{\infty} (-1)^n \frac{an}{\ln n}$

10. $\displaystyle\sum_{n=1}^{\infty} \frac{1}{n^2+3n+2}$

11. $\displaystyle\sum_{n=1}^{\infty} \frac{-3}{(n-3)(n+10)}$

12. $\displaystyle\sum_{n=1}^{\infty} \frac{1}{n^2}$

13. $\displaystyle\sum_{n=1}^{\infty} \frac{1}{n^3}$

14. $\displaystyle\sum_{n=1}^{\infty} \frac{1}{\sqrt[3]{n}}$

10.4 Tests for Convergence

If we run into one of the series described in the last section and we're asked to determine whether or not it converges, we're set. But what happens if we come across a series that we're unfamiliar with and are asked this same question? Well, as you're about to see, there are six tests we can use to figure this out. All we have to do is pick one--and as you can probably guess, it's this picking part that often throws people. Hence, after we get done introducing each of the tests, we're going to provide you with a step-by-step procedure for determining which test to use when you run into an unknown series. Here we go...

---

1. The Simple Divergence Test

Take the limit of a_n as n approaches infinity. If you don't get zero, the series diverges. *NOTE: If you do get zero, no information is obtained--the series may or may not diverge. We have to run further tests to make the determination.*

Examples   10.4.1

1. $\displaystyle\sum_{n=1}^{\infty}\frac{1}{n+1}$

Note that since the limit equals zero here, the test tells us nothing. However, if we recognize that the series is a harmonic series, we can solve using that fact.

$$\lim_{n\to\infty}\frac{1}{n+1}=0 \quad \therefore \text{ This test tells us nothing.}$$

Note: This is a harmonic series, so by definition this
series diverges.

2. $\displaystyle\sum_{n=1}^{\infty}\frac{n^3+5}{n^2+n-1}$

$$\lim_{n\to\infty}\frac{n^3+5}{n^2+n-1}=\lim_{n\to\infty}\frac{3n^2}{2n+1}=\infty$$

$\therefore$ Since the limit doesn't equal zero, the series diverges.

---

2. The Root Test

Given the series, $\displaystyle\sum_{n=1}^{\infty}a_n$

1. Find $\displaystyle\lim_{n\to\infty}\left|a_n\right|^{\frac{1}{n}}=w$
2. If $w>1$, or $w=\infty$, then the series diverges.
3. If $w<1$. the series converges.
4. If $w=1$, the test tells you nothing.

Examples   10.4.2
Use the Root test to determine whether or not the given series converge.

1. $\displaystyle\sum_{n=1}^{\infty}4n^{2n}$

Here we're just following along with the above formula. This problem is straight forward, we get infinity upon solving the limit, hence the series diverges.

(1) $\displaystyle\lim_{n\to\infty}\left(4n^{2n}\right)^{\frac{1}{n}}=\lim_{n\to\infty}\left(4^{\frac{1}{n}}n^2\right)=\infty \quad \therefore \text{ diverges}$

2. $\displaystyle\sum_{n=1}^{\infty} \frac{3^n}{2^{n+3}}$

(1) $\displaystyle\lim_{n\to\infty} \left(\frac{3^n}{2^{n+3}}\right)^{\frac{1}{n}} = \lim_{n\to\infty}\left(\frac{3}{2^{1+\frac{3}{n}}}\right) = \frac{3}{2}$ $\therefore \frac{3}{2} > 1$ $\therefore$ diverges

3. $\displaystyle\sum_{n=1}^{\infty} \frac{5^{n+2}}{n^{2n}}$

(1) $\displaystyle\lim_{n\to\infty} \left(\frac{5^{n+2}}{n^{2n}}\right)^{\frac{1}{n}} = \lim_{n\to\infty}\left(\frac{5^{1+\frac{2}{n}}}{n^2}\right) = \frac{5}{\infty} = 0$ $\therefore$ $0 < 1$ $\therefore$ converges

---

3. The Ratio Test

Given the series, $\displaystyle\sum_{n=1}^{\infty} a_n$

1. Find $\displaystyle\lim_{n\to\infty} \left|\frac{a_{n+1}}{a_n}\right| = w$
2. If $w > 1$, the series diverges.
3. If $w < 1$, the series converges.
4. If $w = 1$, the test tells you nothing.

Examples   10.4.3
Use the Ratio test to determine whether or not the given series converges.

1. $\displaystyle\sum_{b=1}^{\infty} \frac{n!}{2^n}$

(1) $\displaystyle\lim_{n\to\infty} \left|\frac{\dfrac{(n+1)!}{2^{n+1}}}{\dfrac{n!}{2^n}}\right| = \lim_{n\to\infty}\left[\frac{(n+1)!}{2^{n+1}}\left(\frac{2^n}{n!}\right)\right] = \lim_{n\to\infty}\left[\frac{(n+1)n!}{(2^n)2}\left(\frac{2^n}{n!}\right)\right] = \lim_{n\to\infty} \frac{n+1}{2} = \infty$

$\therefore$ the series diverges.

2. $\displaystyle\sum_{b=1}^{\infty} \frac{n^2+1}{3n!}$

(1) $\displaystyle\lim_{n\to\infty} \left| \frac{\left(\dfrac{(n+1)^2+1}{3(n+1)!}\right)}{\dfrac{n^2+1}{3n!}} \right| = \lim_{n\to\infty}\left[\frac{(n+1)^2+1}{3(n+1)!}\left(\frac{3n!}{n^2+1}\right)\right]$

$$= \lim_{n\to\infty} \frac{n^2+2n+2}{(n+1)(n^2+1)} = \lim_{n\to\infty} \frac{n^2+2n+2}{n^3+n+n^2+1} = 0$$

$\therefore$ $0 < 1$ $\therefore$ converges

---

4. The Integral Test

Given the series, $\displaystyle\sum_{n=1}^{\infty} a_n$

If $\displaystyle\int_1^{\infty} a_n\, dn$, converges, then the series converges. If it diverges, the series diverges.
As you can probably infer, the value of this technique is dependent upon the ease of integration.

Examples   10.4.4

1. $\displaystyle\sum_{n=1}^{\infty} \frac{1}{8n+1}$

Very straight forward test. All we do is take the integral of the series. If the integral converges, the series converges, if it doesn't, the series doesn't. Here, we find that the series diverges.

$$\int_1^{\infty} \frac{1}{8n+1}\,dn = \lim_{t\to\infty} \int_1^{t} \frac{1}{8n+1}\,dn$$

$$u = 8n+1, \quad du = 8\, dn$$

$$= \lim_{t\to\infty} \int_1^{t} \frac{1}{u}\left(\frac{du}{8}\right)$$

$$= \lim_{t\to\infty}\left(\frac{1}{8}\ln|u|\right) = \lim_{t\to\infty}\left(\frac{1}{8}\ln|8n+1|\right)\Big|_1^{t} = \lim_{t\to\infty}\left[\left(\frac{1}{8}\ln|8t+1|\right) - \frac{1}{8}\ln|9|\right] = \infty$$

2. $\displaystyle\sum_{n=1}^{\infty} 3ne^{-3n^2}$

Solving this limit, we find that it converges, hence the series converges as well.

$$\int_1^{\infty} 3ne^{-3n^2}\, dn = \lim_{t\to\infty} \int_1^t 3ne^{-3n^2}\, dn$$

$$u = -3n^2, \ du = -6n\, dn$$

$$= \lim_{t\to\infty} \int_1^t \left(3ne^u\right)\left(\frac{du}{-6n}\right)$$

$$= \lim_{t\to\infty} \int_1^t \left(-\frac{1}{2}e^u\right) du$$

$$= \lim_{t\to\infty} \left(-\frac{1}{2}e^u\right)$$

$$= \lim_{t\to\infty} \left(-\frac{1}{2}e^{-3n^2}\right)\Big|_1^t = \lim_{t\to\infty}\left[-\frac{1}{2}e^{-3t^2} + \frac{1}{2}e^{-3(1)^2}\right] = -\frac{1}{2}e^{-\infty} + \frac{1}{2}e^{-3} = \frac{1}{2}e^{-3}$$

---

5. The Limit Comparison Test

Given the series, $\displaystyle\sum_{n=1}^{\infty} a_n$

1. Find another series, $\displaystyle\sum_{n=1}^{\infty} b_n$, that looks similar to a_n and is known by you to either converge or diverge.

2. Take the following limit: $\displaystyle\lim_{n\to\infty} \frac{a_n}{b_n}$. If the limit exists and is greater than zero, then the a_n series converges if the b_n series converges, and it diverges if the b_n series diverges.

3. If the limit that you took in Step #2 does not exist or is not greater than zero, then you b_n was not a good choice. Choose another technique to solve, or start again with a new b_n.

Examples

Use the limit comparison test to determine whether or not the given series converge.

1. $\displaystyle\sum_{n=1}^{\infty} \frac{1996}{(3+5n)^3}$

The whole key to this test is being able to pick another series that you know about, and closely resembles the original series. Usually, the way you do this is by picking one of the types of series we studied in section 10.3-- either a geometric, harmonic, etc...

(1) $a_n = \dfrac{1996}{(3+5n)^3}$, $b_n = \dfrac{1996}{125n^3}$

b_n is a p-series, $3 > 1$ and thus converges.

(2) $\displaystyle\lim_{n\to\infty}\left[\frac{\left(\dfrac{1996}{(3+5n)^3}\right)}{\left(\dfrac{1996}{125n^3}\right)}\right] = \lim_{n\to\infty}\left[\frac{125n^3}{(3+5n)^3}\right]$

$$= \lim_{n\to\infty}\left[\frac{375n^2}{15(3+5n)^2}\right] = \lim_{n\to\infty}\left[\frac{750n}{150(3+5n)}\right] = \lim_{n\to\infty}\left[\frac{750}{750}\right] =$$

$\therefore$ the limit exists, $1 > 0$, $\therefore$ since b_n converges, a_n converges.

2. $\displaystyle\sum_{n=1}^{\infty} \frac{2n^2 - n^{\frac{3}{2}} + 5n^{\frac{1}{3}} + 5}{13n^{\frac{1}{5}} + n^{\frac{1}{2}} + 2n^8}$

This problem looks really bad, but observe how we simplify it:

(1) $a_n = \dfrac{2n^2 - n^{\frac{3}{2}} + 5n^{\frac{1}{3}} + 5}{13n^{\frac{1}{5}} + n^{\frac{1}{2}} + 2n^8}$

Get rid of all but the highest powers in both the numerator and denominator to find a b_n.

$b_n = \dfrac{2n^2}{2n^8} = \dfrac{1}{n^6}$

First, we pick a function that is similar (power wise), yet much less complex.

b_n is a p-series, $6 > 1$ $\therefore$ b_n converges.

$$(2) \quad \lim_{n \to \infty} \left[\frac{\left(\dfrac{2n^2 - n^{\frac{3}{2}} + 5n^{\frac{1}{3}} + 5}{13n^{\frac{1}{5}} + n^{\frac{1}{2}} + 2n^8} \right)}{\left(\dfrac{1}{n^6} \right)} \right] = \lim_{n \to \infty} \left[\frac{2n^8 - n^{\frac{15}{2}} + 5n^{\frac{19}{3}} + 5n^6}{13n^{\frac{1}{5}} + n^{\frac{1}{2}} + 2n^8} \right]$$

$$= \lim_{n \to \infty} \frac{2n^8}{2n^8} = 1$$

Second, when we solve we take advantage of the technique we learned in Chapter 4, section 4.3, to simplify the fraction when dealing with infinite limits. this makes the limit readily solvable.

$\therefore$ The limit exists, $1 > 0$, $\therefore$ since b_n converges, a_n converges.

---

6. Basic Comparison Test

Given the series, $\displaystyle\sum_{n=1}^{\infty} a_n$

1. To detect divergence find another series, $\displaystyle\sum_{n=1}^{\infty} b_n$, that diverges.

   If $b_n \leq a_n$ then $\displaystyle\sum_{n=1}^{\infty} a_n$ diverges too.

2. To detect convergence, find another series, $\displaystyle\sum_{n=1}^{\infty} b_n$, that converges

   If $b_n \geq a_n$ then $\displaystyle\sum_{n=1}^{\infty} a_n$ converges too.

Examples   10.4.6
Use the Basic Comparison test to determine whether or not the given series converges.

1. $\displaystyle\sum_{n=1}^{\infty} \frac{14}{5^n + 2}$

As you can see, this test uses a lot of the same tools the previous tests use, so you shouldn't find this too difficult. Again, the most challenging part is finding a "good" series to use as a comparison.

$a_n = \dfrac{14}{5^n + 2}$ so let $b_n = \dfrac{14}{5^n}$

b_n is a geometric series: $14\left(\dfrac{1}{5}\right)^n$, and since $\dfrac{1}{5} < 1$, b_n converges

Since $b_n > a_n$, a_n also converges.

2. $\displaystyle\sum_{n=1}^{\infty} \frac{14n^2 - 5n - 3n^{\frac{1}{2}}}{19n^{\frac{1}{3}} - n^2 + 28n^5}$

Two significant points that this problem illustrates:

$a_n = \dfrac{14n^2 - 5n - 3n^{\frac{1}{2}}}{19n^{\frac{1}{3}} - n^2 + 28n^5}$ so let $b_n = \dfrac{14n^2}{28n^5} = \dfrac{1}{2n^3}$

b_n is a p - series, $3 > 1$, $\therefore$ it converges.

(1) As a general rule, always create your alternate series from the one your given . Here, we just threw off all but the highest terms in then numerator and denominator. If you look back over the past several examples, you'll see this works quite nicely most of the time.

Is $b_n \geq a_n$?

$\dfrac{1}{2n^3} \geq \dfrac{14n^2 - 5n - 3n^{\frac{1}{2}}}{19n^{\frac{1}{3}} - n^2 + 28n^5}$

$\dfrac{1}{2n^3} - \dfrac{14n^2 - 5n - 3n^{\frac{1}{2}}}{19n^{\frac{1}{3}} - n^2 + 28n^5} \geq 0$

$\dfrac{19n^{\frac{1}{3}} - n^2 + 28n^5 - 28n^5 + 10n^4 + 6n^{\frac{1}{2}}}{38n - 2n^5 + 56n^8} \geq 0$

$\dfrac{10n^4 - n^2 + 6n^{\frac{1}{2}} + 19n^{\frac{1}{3}}}{56n^8 - 2n^5 + 38n} \geq 0$, yes, therefore a_n converges.

(2) It's not always obvious if one series is greater than the other. The best way to figure this out is to do what we did to the left--combine the terms and simplify until you can make an easy determination.

3. $\displaystyle\sum_{n=1}^{\infty} \frac{19^{n-1} + 2}{96}$

$a_n = \dfrac{19^{n-1} + 2}{96}$ so let $b_n = \dfrac{19^{n-1}}{96}$

b_n is a geometric series, $\dfrac{1}{96}(19)^{n-1} > 1$, since $10 > 1$, it diverges

Since $a_n > b_n$, a_n also diverges.

Putting it all together:

You now have a feel for the techniques commonly used in determining convergence. Since there are so many, it will be helpful if you develop a plan for implementing them when faced with an unknown series. We thus provide the following guidelines:

1. Use the Simple Divergence Test first to try to determine divergence right off--you might get lucky.

2. If you see a series with powers of *n* or factorials, try the Root or Ratio tests.

3. If the series appears easy to integrate, use the Integral test.

4. If none of the above apply, use the Limit Comparison test.

5. If not event the Limit Comparison test will work, then give the Basic Comparison test a shot.

Bonus Info:

Do the following when dealing with multiple series within a series, for example:

Given the series, $\sum_{n=1}^{\infty}(a_n + b_n)$

1. If both $\sum_{n=1}^{\infty} a_n$ and $\sum_{n=1}^{\infty} b_n$ converge, then the series converges.
2. If they both diverge, then the series diverges.
3. If one diverges and the other converges, you can't make any simple assessment. You must keep the series together and grind out a solution using one of the prior techniques.

Additional Problems   **10.4**

Use the Simple Divergence test to try to determine if the following series diverge.

1. $\sum_{n=1}^{\infty} \dfrac{1}{8n+1}$ 2. $\sum_{n=1}^{\infty} \dfrac{19^{n-1}+2}{96}$

Use the Integral test to determine if the given series converge.

3. $\displaystyle\sum_{n=1}^{\infty} 2ne^{-4n}$

4. $\displaystyle\sum_{n=1}^{\infty} \frac{32}{8n^2 + 1}$

Use the Limit Comparison test to determine if the given series converge.

5. $\displaystyle\sum_{n=1}^{\infty} \frac{n+1}{n^6 + n^4 + n - 2}$

6. $\displaystyle\sum_{n=1}^{\infty} \frac{32}{(9+n)^{\frac{1}{2}}}$

Use the Root test to determine if the given series converge.

7. $\displaystyle\sum_{n=1}^{\infty} \frac{4n^2}{3^{n+1}}$

8. $\displaystyle\sum_{n=1}^{\infty} \frac{10^n}{2n^n}$

Use the Ratio test to determine if the given series converge.

9. $\displaystyle\sum_{n=1}^{\infty} \frac{(-5)^n}{n!}$

10. $\displaystyle\sum_{n=1}^{\infty} (-1)^n \left(\frac{n^3 + 1}{n^2 + 2} \right)$

Use the Basic Comparison test to determine if the given series converge.

11. $\displaystyle\sum_{n=1}^{\infty} \frac{n^{\frac{1}{2}}}{n - 5}$

12. $\displaystyle\sum_{n=1}^{\infty} \frac{3 + 2^n}{2 + 3^n}$

10.5 Power Series

Remember in the beginning of this chapter when we told you we were going to cross over the bounds of the theoretical into the real world? Wondering when that's going to happen? Well, we're one step away. As promised, in the next section we're going to learn how to approximate functions using the Taylor series. Before we can do this, however, we need to get one final piece of background information in place: We need to learn about *power series*.

So what are power series? Simply series that contain variables in them. Notice that in all the previous sections in which we talked about series, the terms in the series were composed of constants. For example, a typical series we've seen is:

$$\sum_{n=1}^{\infty} \frac{1}{n+2} = \frac{1}{3} + \frac{1}{4} + \frac{1}{5} + \frac{1}{6} + \ldots + \frac{1}{n+2} + \ldots$$

Note that in the expanded series, each term is a constant--no variables. Well in a power series, there is a variable in each of the terms. For example, a typical power series is:

$$\sum_{n=0}^{\infty} \frac{1}{n+2} x^n = \frac{1}{2} + \frac{1}{3}x + \frac{1}{4}x^2 + \frac{1}{5}x^3 + \frac{1}{6}x^4 + ... + \frac{1}{n+2}x^n + ..$$

Formally, we say that a power series is defined as:

$$\sum_{n=0}^{\infty} a_n x^n = a_0 + a_1 x + a_2 x^2 + a_3 x^3 + a_4 x^4 + ... + a_n x^n + ...$$

or, another common form:

$$\sum_{n=1}^{\infty} a_n (x-c)^n = a_0 + a_1(x-c) + a_2(x-c)^2 + a_3(x-c)^3 + ... + a_n(x-c)^n + ...$$

Of special importance with power series is their *interval of convergence*. What do we mean, "interval of convergence"? Well, note that since there are variables (x's) in the power series, some numbers that get plugged into the variable will cause the series to converge while others will cause it to diverge. Typically, a series will have some interval of numbers (as in, $x = 4$ to $x = 20$) for which it will converge--the interval of convergence--which is good to know. Why? You'll see in the next section...

So how do we find the interval of convergence of a power series? We take advantage of the following set of rules:

For power series in the form, $\sum_{n=0}^{\infty} a_n x^n$, one, and only one, of the following statements is true:

1. The power series converges only when $x = 0$.
2. The series is convergent for all values of x.
3. The series is convergent when $|x| < t$ and divergent when $|x| > t$, where t is some positive number.

For power series in the form, $\sum_{n=0}^{\infty} a_n (x-c)^n$, one, and only one, of the following statements is true:

1. The power series converges only when $x = c$.
2. The series is convergent for all values of x.
3. The series is convergent when $|x-c| < t$ and divergent when $|x-c| > t$, where t is some positive number.

Examples   **10.5**

Find the interval of convergence for the given power series.

1. $\displaystyle\sum_{n=0}^{\infty} \frac{1}{n+2} x^n$

The first thing we do in determining the interval of convergence is pick an appropriate test from the last section to perform. In this case, due to the fact that we have a series with powers of n, the Ratio test is a good place to start.

Do the Ratio test:

(1) $\displaystyle\lim_{n\to\infty} \left| \frac{\left(\dfrac{1}{(n+1)+2} x^{n+1} \right)}{\dfrac{1}{n+2} x^n} \right| = \lim_{n\to\infty} \left| \frac{1}{(n+1)+2} x^{n+1} \left(\frac{n+2}{x^n} \right) \right|$

$= \displaystyle\lim_{n\to\infty} \left(\frac{(n+2)}{n+3} \right) |x| = (1)|x| = |x|$

Looking back at the rules from the last section regarding the interpretation of the Ratio test, remember that for numbers greater than 1, the series diverges, for numbers less than 1 the series converges, and when you get 1 it tells you nothing. Here, this translates into the following interval of convergence.

Running the test is not hard, but instead of getting a number like before, now we get a variable in the answer. This is the main difference--we have to interpret the results.

$|x| < 1$, converges: $-1 < x < 1$. $|x| > 1$, diverges: $x > 1$ or $x < -1$

In order to do this interpreting, look back at the Ratio test rules and write them out for the given situation.

To determine exactly what happens at 1 and -1, you plug them into the series and check via direct subsitution:

Plugging in 1:

$\displaystyle\sum_{n=0}^{\infty} \frac{1}{n+2} x^n = \frac{1}{2} + \frac{1}{3} + \frac{1}{4} + ... + \frac{1}{n+2} + ... \Rightarrow$ a harmonic series, $\therefore$ diverges

So 1 is not in the interval of convergence.

Plugging in -1:

$\displaystyle\sum_{n=0}^{\infty} \frac{1}{n+2} x^n = \frac{1}{2} - \frac{1}{3} + \frac{1}{4} + ... + \frac{1}{n+2} x^n + ... \Rightarrow$ an alternating series - -test more:

(1) $\lim\limits_{n \to \infty} \dfrac{1}{n+2} x^n = 0$

(2) $\dfrac{1}{n+1+2} \le \dfrac{1}{n+2}$

$\dfrac{1}{n+3} \le \dfrac{1}{n+2}$ \quad Yes

As you can see, these problems can be tedious, especially when we have to verify the endpoints of the interval of convergence-- it basically results in having to solve a problem within the problem. Don't despair. Just remember the purpose of the exercise and take it one step at a time. All we're doing here is applying what we've learned in the first four sections to a special kind of series--don't make it harder than it is.

$\therefore$ converges at -1.

So, the interval of convergence for this power series is: $-1 \le x < 1$

2. $\displaystyle\sum_{n=0}^{\infty} \dfrac{n^2}{3^{3n}} (x+9)^n$

Essentially we solve this problem just like the last one:

(1) Use the Root test to solve:

First, we decide what test to run to determine convergence or divergence. Looking back at the previous section, the Root test looks like it will do the job--look for series in the past section that resemble the one you're working on here, that's the best way to clue in to the correct test.

$$\lim_{n \to \infty} \left[\dfrac{n^2}{3^{3n}} (x+9)^n \right]^{\frac{1}{n}} = \lim_{n \to \infty} \left[\dfrac{n^{\frac{2}{n}}}{3^3} (x+9) \right] = \dfrac{1}{27}(x+9)$$

The Root test indicates convergence then when:

$-1 < \dfrac{1}{27}(x+9) < 1$

Rewriting this, we get: $\quad -1 < \dfrac{1}{27}x + \dfrac{1}{3} < 1$

$$\left(-\dfrac{4}{3}\right)27 < x < \left(\dfrac{2}{3}\right)27$$

$$-36 < x < 18$$

After doing the Root test, we do some quick algebra based on the test result conditions, thus establishing a preliminary interval of convergence.

What happens at $x = -36$ and $x = 18$?

At $x = -36$: $\displaystyle\sum_{n=0}^{\infty} \dfrac{n^2}{3^{3n}}(-27)^n = \sum_{n=0}^{\infty} \dfrac{n^2}{3^{3n}}(-1^n)(27)^n = \sum_{n=0}^{\infty} (-1)^n \dfrac{27^n n^2}{3^{3n}} \Rightarrow$ alternating series

(i) $\lim\limits_{n\to\infty} \dfrac{27^n n^2}{3^{3n}} = \lim\limits_{n\to\infty} \dfrac{3^{3n} n^2}{3^{3n}} = \lim\limits_{n\to\infty} n^2 = \infty$ $\therefore$ diverges

We say preliminary interval of convergence above because the endpoints need to be tested. Here, at both points we get divergence, so the points are not part of the interval, as we indicate.

At $x = 18$: $\displaystyle\sum_{n=0}^{\infty} \dfrac{n^2}{3^{3n}}(27)^n = \sum_{n=0}^{\infty} \dfrac{n^2}{3^{3n}} 3^{3n} = \sum_{n=0}^{\infty} n^2 \Rightarrow$ diverges

$\therefore$ the interval of convergence is, $-36 < x < 18$

Additional Problems   **10.5**

Find the interval of convergence for the given power series.

1. $\displaystyle\sum_{n=0}^{\infty} \dfrac{2n}{n^2+2} x^n$ 2. $\displaystyle\sum_{n=0}^{\infty} \dfrac{n!}{6^n} x^n$ 3. $\displaystyle\sum_{n=0}^{\infty} \dfrac{n^3}{2^{2n}}(x+25)$

10.6 Taylor Series

This is it! The big moment we've all been waiting for. Hello Real World! We are now going to learn how to approximate functions.

"Halt!" you say. "What do you mean, approximate functions?"

Well, remember Chapter 7, *Funky Functions*, where we learned about transcendental functions--i.e., nonalgebraic functions like sine and cosine? It turns out that these types of functions pop up all the time in real world problems. Unfortunately, as you know, transcendental functions are not all that easy to work with. Hence, when you're trying to solve a problem with one or more of these types of functions in it, and super accuracy is not required, having to deal with them is a real drag. Why? Because the amount of effort which you have to expend to solve the problem is tremendous. Thus, in these types of situations it's often desirable to approximate the transcendental function--that is, convert it to an algebraic function that's much easier to work with.

What happens to the accuracy of the answer when you do this? Well, it's a little off. But, as you're about to see, you can control how off the answer is because during the course of the conversion you get to determine to how many decimals of accuracy you need. Thus, the only thing you really lose when you use a Taylor series to convert a transcendental function to an algebraic one is a big headache. Engineers use these this technique all the time.

Now that we've wet your appetite, it's time to tell you exactly what a Taylor series is:

A Taylor series is a power series in the form:

$$\sum_{n=0}^{\infty} \frac{f^{(n)}(c)}{n!}(x-c)^n$$

which, when $c = 0$, simplifies to,

$$\sum_{n=0}^{\infty} \frac{f^{(n)}(0)}{n!}(x)^n$$

Note: The $f^{(n)}(c)$ means this--the n refers to the derivative order, and the (c)
refers to what number gets plugged into the resulting function. Example,
$f(x) = \cos x$, then for the $n = 3$ term given $c = \pi$, we'd have: $f'''(\pi) = \sin(\pi)$.

Important point: When we translate a function to a Taylor series, the approximation is only valid over the series' interval of convergence. Unfortunately, determining the interval of convergence of a Taylor series is a highly nontrivial task. Fortunately, though your textbook and lectures may briefly mention a method for determining such intervals, in first year calculus you aren't going to be asked to do this--so don't worry about it. Below we list the intervals of convergence of the some common Taylor series. Be aware of them.

Taylor series	Interval of Convergence
1. $\sin x = \sum_{n=0}^{\infty} (-1)^n \frac{1}{(2n+1)!} x^{2n+1}$	$(-\infty, \infty)$
2. $\cos x = \sum_{n=0}^{\infty} (-1)^n \frac{1}{(2n)!} x^{2n}$	$(-\infty, \infty)$
3. $e^x = \sum_{n=0}^{\infty} \frac{x^n}{n!}$	$(-\infty, \infty)$
4. $\ln(x+1) = \sum_{n=0}^{\infty} (-1)^n \frac{1}{n+1} x^{n+1}$	$(-1, 1]$

That's a lot of information. Don't worry. Though the definitions above look complex, they're not hard to work with at all The following examples show you how to put it all together--how to translate a function to Taylor series form with minimal pain.

Examples

Find the Taylor series for the following functions.

1. $f(x) = \sin x$

As you can see from the above function, there is no c, hence $c = 0$, so we use the second Taylor series form as follows:

$$\sin x = 0 + \frac{\cos 0}{1!}x + \frac{-\sin 0}{2!}x^2 + \frac{-\cos 0}{3!}x^3 + \frac{\sin 0}{4!}x^4 + \frac{\cos 0}{5!}x^5 + \ldots$$

Rewriting to see what we've got:

$$\sin x = x + 0 - \frac{1}{3!}x^3 + 0 + \frac{1}{5!}x^5 + \ldots = x - \frac{1}{3!}x^3 + \frac{1}{5!}x^5 + \ldots$$

This resulting series looks suspiciously like an alternating series. In fact, if you write out a few more terms and look at it a while, you'll see that $\sin x$ can be written as the following series:

$$\sin x = x - \frac{1}{3!}x^3 + \frac{1}{5!}x^5 - \frac{1}{7!}x^7 + \frac{1}{9!}x^9 + \ldots + (-1)^k \frac{1}{(2k+1)!}x^k + \ldots$$

which is: $\displaystyle\sum_{k=0}^{\infty}(-1)^k \frac{1}{(2k+1)!}x^k$

2. $f(x) = e^{-2x}$

As you can see, these problems are not that bad. Essentially, if you know how to take higher order derivatives, you're set. All we do here is follow the formula, take the derivatives, and plug in 0 since $c = 0$. Then we simplify.

$$e^{-2x} = \frac{e^{-2x}}{0!} + \frac{-2e^{-2x}}{1!}x + \frac{4e^{-2x}}{2!}x^2 + \frac{-8e^{-2x}}{3!}x^3 + \ldots$$

$$= \frac{e^{-2(0)}}{1} + \frac{-2e^{-2(0)}}{1!}x + \frac{4e^{-2(0)}}{2!}x^2 + \frac{-8e^{-2(0)}}{3!}x^3 + \ldots$$

$$= \frac{1}{1} + \frac{-2}{1!}x + \frac{4}{2!}x^2 + \frac{-8}{3!}x^3 + \ldots$$

$$= 1 - 2x + \frac{4}{2!}x^2 + \frac{-8}{3!}x^3 + \ldots$$

Writing out a few more terms, we find that the series is:

$$e^{-2x} = 1 - 2x + \frac{4}{2!}x^2 + \frac{-8}{3!}x^3 + \frac{16}{4!}x^4 + \frac{-32}{5!}x^5 + \ldots + (-1)^{n+1}\frac{2^n}{n!}x^n + \ldots$$

3. $f(x) = \cos x,$ $\qquad c = \pi$

Since c is specified, we need to use the first formula:

$$\cos x = \frac{\cos x}{0!}(x-\pi)^0 + \frac{-\sin x}{1!}(x-\pi)^1 + \frac{-\cos x}{2!}(x-\pi)^2 + \frac{\sin x}{3!}(x-\pi)^3$$

$$+ \frac{\cos x}{4!}(x-\pi)^4 + ...$$

$$\cos x = -1(x-\pi) + 0 + \frac{1}{2!}(x-\pi)^2 + 0 - \frac{1}{4!}(x-\pi)^4 + ..$$

$$\cos x = -(x-\pi) + \frac{1}{2!}(x-\pi)^2 - \frac{1}{4!}(x-\pi)^4 + ...$$

Hence, the Taylor sereis is:

$$\cos x = \sum_{k=0}^{\infty} \frac{1}{2k!}(x-\pi)^k = -(x-\pi) + \frac{1}{2!}(x-\pi)^2 - \frac{1}{4!}(x-\pi)^4$$

$$+ \frac{1}{6!}(x-\pi)^6 + ... + \frac{1}{2k!}(x-\pi)^k + ...$$

4. Approximate, $\int_0^1 \cos x^2 dx$, to four decimal places.

Basically, what we're looking to do here is convert this transcendental function to an algebraic one that's easier to integrate. Hence, we need to translate it to Taylor series format. Note that substitution is legal...

The Taylor series for $\cos x^2 = ?$

Step 1, make life easy, let $u = x^2$

$$\cos u = \frac{\cos u}{0!}u^0 + \frac{-\sin u}{1!}u^1 + \frac{-\cos u}{2!}u^2 + \frac{\sin u}{3!}u^3 + \frac{\cos u}{4!}u^4 + ...$$

In writing out the series, don't go wild. Just write out the first 4 terms. Why? We want 4 decimal places.

$$\cos u = 1 - 0 - \frac{1}{2!}u^2 + 0 + \frac{1}{4!}u^4 + ...$$

$$\cos u = 1 - \frac{1}{2!}u^2 + \frac{1}{4!}u^4 - \frac{1}{8!}u^6 + ...$$

So, $\cos x^2 = 1 - \frac{1}{2!}(x^2)^2 + \frac{1}{4!}(x^2)^4 - \frac{1}{8!}(x^2)^6 = 1 - \frac{1}{2!}x^4 + \frac{1}{4!}x^8 - \frac{1}{8!}x^{12}$

$$\therefore \int_0^1 \cos x^2\, dx \approx \int_0^1 \left(1 - \frac{1}{2!}x^4 + \frac{1}{4!}x^8 - \frac{1}{8!}x^{12}\right)dx$$

$$= \int_0^1 dx + \int_0^1 \left(-\frac{1}{2!}x^4\right)dx + \int_0^1 \left(\frac{1}{4!}x^8\right)dx + \int_0^1 \left(-\frac{1}{8!}x^{12}\right)dx$$

$$= x\Big|_0^1 + \left(-\frac{1}{10}x^5\right)\Big|_0^1 + \left(\frac{1}{216}x^9\right)\Big|_0^1 + \left(-\frac{1}{524{,}160}x^{13}\right)\Big|_0^1$$

$$= 1 - \frac{1}{10} + \frac{1}{216} - \frac{1}{524{,}160} = 0.9046$$

Additional Problems   10.6

Find the Taylor series for the following problems.

1. $f(x) = e^x$ 2. $f(x) = \cosh x$

3. Approximate, $\int_0^1 \sin x^3\, dx$, to five decimal places.

Bonus Info: Undoubtedly you haven't realized this yet, but in the real world we're rarely given functions to help us solve problems--we have to come up them ourselves. How do we do this? Well, one of the most basic ways is to collect data on something we're interested in and then do what's known as *curve fitting* to come up with a function to represent it.

For example, lets say we're testing the contraction force of an artificial heart. In doing this, we record the force output as a function of time, coming up with the relationship, $FO = f(t)$, where FO is the dependent variable representing force output, t is the independent variable representing time, and f is the function relating the two groups of numbers. If we collect enough data, clearly we can plot the values we obtain for both t and FO on a Cartesian coordinate system. Then in order to figure out what the function $f(t)$ is that relates these two groups of numbers, all we have to do is apply some type of curve fitting to the graph.

How is this done? Well, we simply designate how many terms we want in our function-- think back to the Taylor series section when we approximated the integrals and we specified how accurate we wanted to be--and then let a computer crank through the numbers using some technique to generate an algebraic function (Note: Don't worry about specific curve fitting techniques, that's not important). After coming up with the function, the function is plotted on the same graph as the data to determine how accurately it mimics the data--in other words, how good the fit is. If we decide the fit is not good enough, we simply increase the number of terms in the function and repeat the process.

Chapter 11

Vectors

11.1 Vectors & Scalars--The Basic Idea
11.2 Unit Vectors & Components of Vectors
11.3 Vector Operations

Believe it or not, we're not going to see any calculus in this chapter. Rather, we're going to learn about a special kind of mathematical variable known as the vector. Why are we doing this? Because learning about vectors sets us up nicely for the next chapter, where we'll take a look at functions composed of vectors--known as vector functions--and apply the calculus we've learned on them. Why are vector functions significant? Because we see them everywhere in science. In short, they're a very real world application.

Now don't skip this chapter! We know you've probably seen this material before in your science classes--probably multiple times by now--but that doesn't matter. In order to succeed in the next chapter you need to have a good understanding of vectors and it needs to be fresh in your mind. This chapter sets you up perfectly for what is to come. Take the time to read it. You won't be disappointed.

11.1 Vectors & Scalars--The Basic Idea

Imagine this scene: You're an air traffic controller at Chicago's O'Hare airport on a cold rainy night in the middle of November. Suddenly, a distress signal comes racing in through your headset: It's a small twin engine plane, three miles out. The pilot is lost in the fog, running out of fuel, and beginning to panic. As he frantically asks you for the location of the nearest runway, do you:

(a) Tell him the distance he is from the runway.
(b) The direction he must pursue to get to the runway.
(c) All of the above.

Let's think about this for a minute. If you choose (a) or (b) what happens? That's right, (a) leaves the pilot guessing which way to go and (b) leaves him guessing how far to go. Not good. Hence (c) is the correct choice because both pieces of information need to be transmitted to the pilot in order for him to land safely. One way to transmit this information is via a vector.

So what exactly is a vector? *A vector is simply a special kind of mathematical variable used to represent things that require both a magnitude (in other words, a quantity) and a direction to be completely described.* Things like velocity, acceleration, force, and displacement are examples of things represented by vectors. In our above example, the pilot needed to know how and in which direction to go to get to the runway--i.e., a displacement. See Fig. 11.1.

A couple of things to note: First, due to the fact that vectors specify direction, they're useless unless they're defined with respect to some coordinate system. For example, in the above situation, if you told the pilot to go 3 miles at 30 degrees, the pilot wouldn't have the foggiest idea of what you were talking about. Why? Because the 30 degrees, by itself, is meaningless. You have to specify direction with respect to a coordinate system--as in, go 3 miles at 30 degrees north of west. That's meaningful because the direction is made with respect to a standard coordinate system known to both parties.

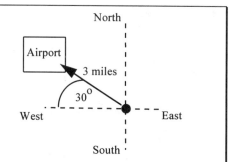

Fig. 11.1: A diagram showing the pilot's situation: The pilot's plane is represented by the black dot. The pilot needs to get to the airport, hence he needs to go 30 degrees north of west for 3 miles.

Second, people get the term *magnitude* confused all the time. Magnitude is just another term for an *amount* or *quantity*. If we say the person ran three miles north, the amount of distance, or magnitude, is three miles, the direction is north. Also, you'll often hear people throw terms around like, *velocity vector*, *displacement vector*, or *force vector*. Don't let these terms fool you. They're just specifying what a particular vector represents--a velocity vector is a vector that represents velocity, a displacement vector is a vector that represents a displacement, etc.

This is quite a bit more complicated than what we're use to dealing with in terms of variables, yes? That's because the variables we've been using up until now are known as *scalars*--i.e., a fancy name for a mathematical variable used to represent things that require only a magnitude to be completely defined. Things like speed, the number of people in attendance at a baseball game, or the distance the pilot was from the runway are examples of things represented by scalars.

As far as notation goes, vectors are represented pictorially as straight lines. They have an arrow on the end indicating which way they're pointing, they're oriented in their appropriate direction, and their length is equal, proportionally, to their magnitude. The end of the vector with the arrow is called the head, and the other end is called the tail. Due to the fact that vectors are meaningless unless they're defined in terms of a coordinate system, we specify their tail as being located at the origin of their coordinate system. See Fig. 11.2.

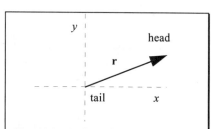

Fig. 11.2: A pictoral representation of a vector **r**. Note that the tail of the vector is always rooted at the origin of a coordinate system--in this case, a Cartesian coordinate system.

Numerically, vectors are represented in this book as letters in bold print. Scalars are represented as they have been throughout the book--letters in italic print.

11.2 Unit Vectors & Components of Vectors

As we've said, specifying a vector with respect to some coordinate system is absolutely essential if the vector is to have any meaning. One way to do this, as we've shown above, is to spell out the direction of a vector relative to the earth's north-south-east-west coordinates. While this works for some applications, it's useless for others because such a coordinate system is irrelevant for them. For example, if you want to use vectors to describe the position of a projectile fired up into the air, north-south-east-west directions won't help you because the projectile, besides going in one of the north-south-east-west directions, is also going up. What's "up" in the north-south-east-west coordinate system? It doesn't exist. Hence that coordinate system can't be used.

So, in order to solve this problem, we commonly use a standard coordinate system that's more flexible--e.g., the Cartesian coordinate system--and what are known as *unit vectors* to help us navigate in the system. We all know what a Cartesian coordinate system is, but what is unit vector?

Well, *a unit vector is simply a vector of magnitude one*. In other words, a vector that's only one unit long. Their purpose is directional--they help us specify the direction of any vector in the coordinate system. In the case of the Cartesian coordinate system, there are three such unit vectors: **i, j,** and **k**. As you can see from Fig. 11.3, each of the unit vectors points in the positive direction of one of the three axes. By convention, **i** is the unit vector in the positive *x* direction, **j** is a unit vector in the positive *y* direction, and **k** is a unit vector in the positive *z* direction.

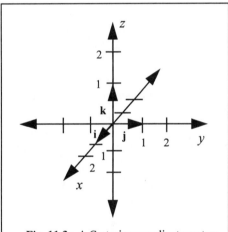

Fig. 11.3: A Cartesian coordinate system with the unit vectors **i, j,** and **k** shown.

This is nice, but how exactly does it help us specify the direction of a vector in the Cartesian coordinate system? Well, if we happen to have a vector 5 units long moving in the positive *x*-axis direction, we can simply write it as 5**i**. Why? Because the 5 tells us the vector's magnitude and the **i** tells us its direction. Similarly, if we have a vector 3 units long running along the positive *y*-axis, we simply write it as, 3**j**. If, however, we have a vector 4 units long running along the negative *y*-axis, we write 4(-**j**), or simply -4**j**. Not too difficult, right?

But what happens if we have a vector pointing out in some direction other than along one of the axes, as in Fig. 11.4? (For simplicity, we'll illustrate only 2-D vectors here). In these instances, we have to figure out what combination of vectors in the **i**, and **j** directions (and **k** if we were dealing with a third dimension) could be added together to get the vector in question. For example, in Fig. 11.5 we have a vector four units long in the positive *x*-direction added to a vector 3 units long in the positive *y*-direction, resulting in a vector which is written as, 4**i** + 3**j**.

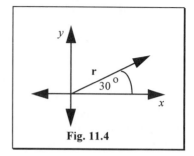

Fig. 11.4

How do we go about figuring out these combinations though? We use trigonometry. For example, let's say we're told we have a vector 5 units long at 30 degrees counter clockwise from the positive axis. In order to figure out what combination of vectors in the **i** and **j** directions could be added together to get this vector, we draw a right triangle, with the hypotenuse being the vector, the vertical side equaling the vector in the **j** direction, and the horizontal side equaling the vector in the **i** direction. Then we just use basic trigonometric relations to solve for the sides. See Fig. 11.6.

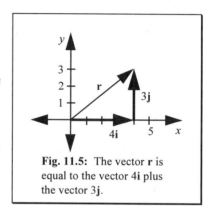

Fig. 11.5: The vector **r** is equal to the vector 4**i** plus the vector 3**j**.

The step-by-step method below summarizes these steps, and the examples below illustrate exactly how to complete them. Make sure you practice this a lot. You're going to be seeing this notation throughout you academic careers and you'll save yourself a lot of headaches if it becomes second nature to you now.

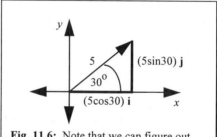

Fig. 11.6: Note that we can figure out what this vector's **i** and **j** components are by drawing a right triangle and using some basic trig.

Method for breaking a vector into its component parts:

1. Define a coordinate system.

2. Sketch the vector in the coordinate system.

3. Use basic trigonometry to complete the transformation by constructing and using a right triangle with hypotenuse equal to the vector, opposite side equal to the vector in the **j** direction, and adjacent side equal to the vector in the **i** direction.

Examples   **11.2.1**

Note: All directions are specified as angles starting from the positive *x*-axis and going counter-clockwise.

1. Given the vector, **a**, magnitude = 10, direction = 23°

   Break the vector into its component parts.

Step 1: Define a coordinate system.

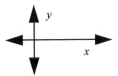

The first step is often already done for you as most problems specify a coordinate system. If you run into one that doesn't, however, don't forget to do this--it's central to giving your solution meaning.

Step 2: Sketch the vector.

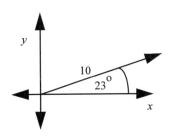

This step is very straight forward. Note that you don't have to be exact when sketching the vector--the whole idea is to just get a rough idea of where it is with respect to the coordinate system. Labeling the magnitude of the vector as well as the angle help.

Step 3: Use basic trigonometry to complete the transformation.

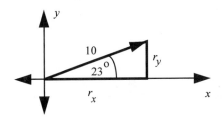

This is the critical step. In order to be successful what you have to do is visualize a triangle forming as we've done here. The triangle's vertical side corresponds to the magnitude of the vector's y component, and the triangle's horizontal side corresponds to the magnitude of the vector's x component.

$$r_x = 10\cos 23 = 9.2$$
$$r_y = 10\sin 23 = 3.9$$

$$\mathbf{a} = 9.2\mathbf{i} + 3.9\mathbf{j}$$

After determining the magnitudes, we have to specify the vector's direction (this is where the unit vectors come in). To do this you match the magnitudes with their corresponding unit vectors, multiply the unit vectors and their magnitudes together, and then add the two resulting quantities together. (Note: Since the unit vectors have themselves a magnitude of 1, they don't change the magnitude).

*The result, as shown above, is a vector represented in terms of its component parts. Remember, **i** is the unit vector in the positive x direction, **j** in the positive y.*

2. Given the vector **t**, magnitude = 317, direction = -103°

   Break the vector into its component parts.

Step 1: Define a coordinate system.

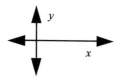

Just like in the last example. You may think this to be a trivial step, but it's not. You've got to specify your coordinate system or your answer will be meaningless. For example, what if we skipped this step and you assumed y was on the horizontal axis instead of the vertical like we chose? You'd be pretty confused, yes?

Step 2: Sketch the vector.

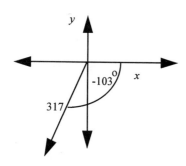

*In sketching this vector immediately note that we're in the third quadrant--in other words x and y values are both negative, hence **i** and **j** are both negative since we're pointing in the negative x and negative y directions.*

Step 3: Use basic trigonometry to
         complete the transformation.

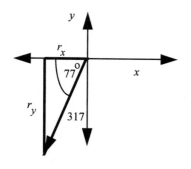

Same idea as in the last example, get that triangle visualized, compute the magnitudes, then match them up with their directions and add. Don't forget that this time both unit vectors are negative as we explained above. This comes out in the answer as shown below.

Note that the only real complication here is having to calculate the angle in the triangle. If you remember that a semicircle is 180 degrees, this is no real problem.

$$108° - 103° = 77°$$

$$r_x = 317\cos 77 = 71.3$$
$$r_y = 317\sin 77 = 309$$

recall that both unit vectors are negative,

thus: $t = -71.3\mathbf{i} - 309\mathbf{j}$

3. Given the vector **a**, magnitude = 21, direction = 159°

   Break the vector into its component parts.

Step 1: Define a coordinate system.

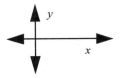

Step 2: Sketch a vector.

*This vector falls in the second quadrant--in other words x values are negative and y values are positive. Hence, **i** is negative here, while **j** is positive.*

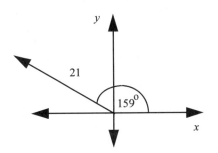

Step 3: Use basic trigonometry to complete the transformation.

*By now you should be use to creating and using the triangle that pops up in these types of problems. Again, the triangle's vertical side corresponds to the magnitude of the vector's y component, while the triangle's horizontal side corresponds to the magnitude of the vector's x component. After we determine the magnitudes, we multiply each by the appropriate unit vector (don't forget here **i** is negative), and then add the results together.*

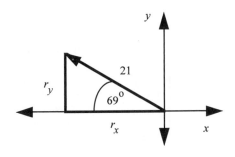

$180° - 159° = 69°$

$r_x = 21\cos 69 = 7.5$

$r_y = 21\sin 69 = 19.6$

$\Rightarrow$ -**i**, +**j**, $\therefore$ **a** = -7.5**i** + 19.6**j**

On some occasions you may find it desirable to convert a vector from its component form to regular form--that is, simply specifying a magnitude and direction. This is done as follows:

Converting a vector from component to regular form:

Given the vector, $\mathbf{r} = r_x \mathbf{i} + r_y \mathbf{j}$

(1) The magnitude, $|\mathbf{r}| = r = \sqrt{r_x^2 + r_y^2}$

(2) The direction, $\theta = \tan^{-1}\left(\dfrac{r_y}{r_x}\right)$

Examples   11.2.2

Convert the vectors in component form back to regular vector form.

1. $\mathbf{a} = 2\mathbf{i} + 3\mathbf{j}$

$|\mathbf{a}| = a = \sqrt{(2)^2 + (3)^2} = \sqrt{13}$

$\theta = \tan^{-1}\dfrac{3}{2} = 56.3°$

$\therefore \ \mathbf{a} = \sqrt{13}, \ 56.3°$

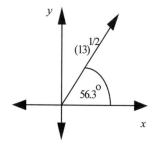

*Note that the angle calculated is always between the vector and the x-axis and that it's measured in the quadrant that the vector lies. How do you know what quadrant the vector is in? Look at the unit vectors. Here **i** and **j** are both positive, thus the vector is in the first quadrant. Check out the graph.*

2. $\mathbf{t} = -173\mathbf{i} - 12\mathbf{j}$

$|\mathbf{t}| = t = \sqrt{(-173)^2 + (-12)^2}$

$\quad = 173.4$

$\theta = \tan^{-1}\left(\dfrac{12}{173}\right) = 3.97°$

$\theta = 180° + 3.97° = 183.97°$

or

$\theta = 3.97° - 180° = -176.03°$

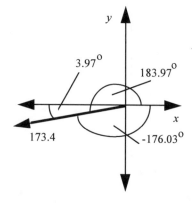

Here the vector is in the third quadrant--both i and j are negative. Hence, when we calculate the angle it appears as is shown to the left, however, we usually express the angle of the vector in our final answer with respect to positive i, hence we have to add or subtract 180 degrees.

$\mathbf{t} = 173.4, \ 183.97°$ or $\mathbf{t} = 173.4, \ -176.03°$

3. $\mathbf{r} = 5\mathbf{j}$

$|\mathbf{r}| = r = \sqrt{(5)^2} = 5$

$\theta = 90°$

$\mathbf{t} = 5, \; 90°$

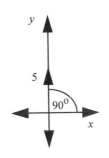

In this problem we only have one component of the vector, hence you have to use a little common sense when calculating the vector--actually you don't calculate, you just look. Since the vector is just going in the positive $\mathbf{j}$ direction, it's obviously 90 degrees away from the positive $\mathbf{i}$ direction.

Additional Problems   11.2

Break the given vectors into their component form. Note: All directions are specified as angles starting from the positive *x*-axis and going counter clockwise.

1. $r = 30, \; \theta = 30°$

2. $r = 5, \; \theta = 150°$

3. $r = 2, \; \theta = -30°$

4. $r = 6, \; \theta = 194°$

Convert the given vectors to regular form.

5. $\mathbf{r} = 2\mathbf{i} + 5\mathbf{j}$

6. $\mathbf{r} = -4\mathbf{i} + 67\mathbf{j}$

7. $\mathbf{r} = 2\mathbf{i} - 6\mathbf{j}$

8. $\mathbf{r} = -2\mathbf{i} - 5\mathbf{j}$

11.3 Vector Operations

Now that we know what vectors are as well as how to express them mathematically, we're ready to learn how to do some basic mathematical operations on them. Specifically, in this section we're going to learn how to:

1. Add and subtract vectors.
2. Multiply scalars and vectors together to modify the magnitude of the vector.
3. Multiply two vectors together to get a scalar using the *dot product*.
4. Perform the *cross product* on two vectors to obtain a new vector.

We begin with the first and simplest of the four operations: Addition/subtraction of vectors. In order to add or subtract two vectors, all we have to do is break them into their component parts, then add the corresponding components together. In other words, the **i**'s get added together, the **j**'s get added together, and in the case of three dimensions, the **k**'s get added together. This is summarized below:

Adding/Subtracting Vectors:

1. Break the vectors in question down into their component parts.

2. Add the components together. The normal rules for addition apply.

Examples   11.3.1

Add the given vectors together.

1. $a = 2i + 7j$

    $b = 3i + 4j$

    In this problem, and in the rest of these for that matter, we'll already have the vectors broken down into their component form. All we need to do, then, is add the components together. Just remember, the i's get added together, the j's together, and if there are any, the k's together.

    $a + b = 2i + 7j + 3i + 4j$

    $\quad = 5i + 11j$

2. $a = 10i - 2j$

    $b = -12i + 5j$

    Nothing difficult here. The only difference is a few negative values. Pay attention to your signs and you'll have no problems.

    $a + b = 10i - 2j + (-12i + 5j)$

    $\quad = -2i + 3j$

3. $a = i - 2j + 10k$

    $b = -i + 15j$

    This problem has a couple of subtle points that are worth noting.

    First, when you see a negative sign alone in front of one of the unit vectors, or simply a unit vector standing alone, there's actually a 1 sitting out in front of it--it's just not shown, it's understood.

    $a + b = (i - 2j + 10k) + (-i + 15j)$

    $\quad = (1i - 2j + 10k) + (-1i + 15j + 0k)$

    $\quad = 0i + 13j + 10k = 13j + 10k$

    Second, the vectors are expressed in 3-dimensions, but not all components are shown in a vector, the components not shown are equal to zero.

Subtract the following vectors.

4. $a = 5i - 6j + 2k$

    $b = -4i + 2j + 8k$

    Don't let the subtraction fool you. Just keep the signs straight--this is not rocket science.

    $a - b = (5i - 6j + 2k) - (-4i + 2j + 8k)$

    $\quad = 5i - 6j + 2k + 4i - 2j - 8k$

    $\quad = 9i - 8j - 6k$

5. $\mathbf{a} = 5\mathbf{j}$

   $\mathbf{b} = -3\mathbf{i} + 6\mathbf{k}$

Note: When you're actually solving these problems you don't have to write out that step where we throw in the zeroes--it's understood. We just did it for emphasis.

$\mathbf{a} - \mathbf{b} = (5\mathbf{j}) - (-3\mathbf{i} + 6\mathbf{k})$

     $= (0\mathbf{i} + 5\mathbf{j} + 0\mathbf{k}) - (-3\mathbf{i} + 0\mathbf{j} + 6\mathbf{k})$

     $= 3\mathbf{i} + 5\mathbf{j} - 6\mathbf{k}$

Multiplying a vector and a scalar together is even easier:

> *Method for multiplying a vector by a scalar to modify the magnitude of the vector:*
>
> 1. If the vector is in component form, simply multiply each component by the scalar. If the vector is in regular form, multiply the magnitude of the vector by the scalar.

Examples   **11.3.2**

Multiply the given vectors and scalars.

1. $\mathbf{a} = 2\mathbf{i} + 3\mathbf{j}, \; s = 5$

This is really easy, just basic multiplication. All we're doing is multiplying each component of the vector by the scalar.

$s\mathbf{a} = 5(2\mathbf{i} + 3\mathbf{j}) = 10\mathbf{i} + 15\mathbf{j}$

2. $\mathbf{a} = -5\mathbf{i} + \mathbf{j} - \mathbf{k}, \; s = -3$

Same thing, different vector and scalar. Noting complicated here.

$s\mathbf{a} = -3(-5\mathbf{i} + \mathbf{j} - \mathbf{k}) = 15\mathbf{i} - 3\mathbf{j} + 3\mathbf{k}$

Now things begin to get a tad hairy. We're going to multiply two vectors together in such a way as to yield a scalar. This scalar tells us the magnitude of the first times the projection of the second onto the first. By projection we mean the magnitude the second vector has in the direction of the first. How do we do this? We use what is known as the *dot product* (sometimes called the *scalar product*), which is defined as follows:

> *Dot product:*
>
> Given the vectors $\mathbf{a}$ and $\mathbf{b}$, with angle θ between them,
>
> $\mathbf{a} \cdot \mathbf{b} = |\mathbf{a}||\mathbf{b}|\cos\theta$

Which is really great if someone takes the time to give you two vectors and tells you the angle between them. But what if we're given two vectors in component form? Simply follow these steps:

Method for multiplying a vector by a vector to get a dot (scalar product):

1. Multiply the vectors' corresponding components--i.e., multiply the **i**'s together, the **j**'s together, the **k**'s together.

2. Add the results.

For example, given: $\mathbf{r} = a\mathbf{i} + b\mathbf{j} + c\mathbf{k}$, $\mathbf{t} = d\mathbf{i} + e\mathbf{j} + f\mathbf{k}$,

$\mathbf{r} \cdot \mathbf{t} = ad + be + cf$

CAUTION: The vector $\mathbf{r} = 2\mathbf{i} + 0\mathbf{j}$ can be written as $\mathbf{r} = 2\mathbf{i}$. If you are using the dot product on a vector like this, don't forget that it has a zero component, and don't forget to include that component in your calculation.

Examples   **11.3.3**
Find the scalar products of the given vectors.

1. $\mathbf{a} = 3\mathbf{i} + 2\mathbf{j}$, $\mathbf{b} = 7\mathbf{i} + 5\mathbf{j}$

Once you get the steps down, this is really easy. Remember, all we're doing here is multiplying the corresponding components together and then adding the results. As you can see, we end up with a scalar.

$\mathbf{a} \cdot \mathbf{b} = (3\mathbf{i} + 2\mathbf{j}) \cdot (7\mathbf{i} + 5\mathbf{j})$

$= (3)(7) + (2)(5) = 31$

2. $\mathbf{a} = 2\mathbf{i}$, $\mathbf{b} = \mathbf{i} + 8\mathbf{j}$

The only thing that can make these problems difficult is if you forget to include the zero components when you multiply. Here, vector a has no j component, it's zero. Don't forget to include it.

$\mathbf{a} \cdot \mathbf{b} = (2\mathbf{i}) \cdot (\mathbf{i} + 8\mathbf{j})$

$= (2\mathbf{i} + 0\mathbf{j}) \cdot (\mathbf{i} + 8\mathbf{j})$

$= (2)(1) + (0)(8) = 2$

Finally we come to the cross product. The cross product, sometimes called the vector product, involves multiplying two vectors together in such a way as to yield a new vector. It's magnitude is defined as follows:

Cross product:

Given the vectors **a** and **b**, with angle θ between them,

$$|\mathbf{a} \times \mathbf{b}| = |\mathbf{a}||\mathbf{b}|\sin\theta$$

What about its direction? Well, the direction of the new vector is perpendicular to the plane formed by vectors **a** and **b**. Note, however, that this vector can be pointing out on either side of the plane. We figure out which way it's pointing by using what's known as the *right hand rule*. This is best seen by example, so observe:

As you can see from Fig. 11.7 and Fig. 11.8, the direction of the new vector formed as a result of the cross product depends on the position of **a** and **b** on the plane, as well as the order in which the cross product is taken.

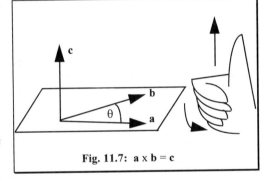

Fig. 11.7: a x b = c

In order to quickly see which way the resulting vector goes, use the right hand rule as shown in the figures. In Fig. 11.7, **a** is crossed into **b**, with the direction of rotation being counter-clockwise. Hence, take your right hand and position it such that the fingers move counter-clockwise when curled. The direction of the thumb is the direction of the new vector.

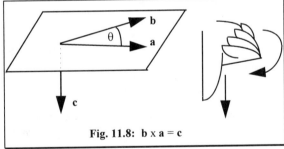

Fig. 11.8: b x a = c

In Fig. 11.8, **b** is crossed into **a** such that the motion is clockwise. Curl the fingers in your right hand in a clockwise direction and presto, the direction of the thumb is the direction of the new vector.

Again, this is all great if we're given **a** and **b** and θ, but what if we're given the two vectors in component form? Well, there is a hard way to do this and an easy way. Since both work equally well, we'll skip right to the easy way.

Method for performing the cross product on two vectors:

1. Remember, order is everything. (**a** x **b**) is not the same as (**b** x **a**).

2. Write down on a corner of your paper the following short diagram:

3. Multiply each component of the first vector by each component of the second vector. Look at the above diagram to figure out the resulting direction.

Example: When you multiply an **i** by a **j**, you get a **k**. When you multiply a **j** by a **k**, you get an **i**, and when you multiply a **k** by an **i** you get a **j**. When you go the other way--i.e., counterclockwise, multiplying an **i** by a **k**, you get a -**j**, a **k** by a **j**, you get a -**i**, a **j** by an **i**, you get a -**k**. Also, when you cross like terms--i.e., **i** and **i**, you get zero because the sine of zero is zero. (Remember the definition of cross product: **a** x **b** = $ab\sin\theta$.

Examples

11.3.3

Find the cross products of the following vectors.

1. **a** $= 3\mathbf{i} + 2\mathbf{j}$, **b** $= 7\mathbf{i} + 5\mathbf{j}$

Remember to use our diagram, shown here to the right. When moving clockwise, you get a positive result, counterclockwise a negative result. In this example, we cross 3i into 5j

$$\mathbf{a} \times \mathbf{b} = (3\mathbf{i} + 2\mathbf{j}) \times (7\mathbf{i} + 5\mathbf{j})$$

$$= 15\mathbf{k} - 14\mathbf{k} = \mathbf{k}$$

which results in a positive 15k. We then cross 2j into 7i, which results in -14k. Also, remember that when like components are crossed, the results are zero. Hence, 3i cross 7i is zero and 2j cross 5j is zero.

2. **a** $= 2\mathbf{i} + 5\mathbf{j} + 10\mathbf{k}$, **b** $= 7\mathbf{i} - 5\mathbf{j} - 3\mathbf{k}$

As you can see, the more terms we get, the more complex the cross product becomes. Once you get use to using the diagram above, you can normally solve problems in one step as shown to the left. The thought process is shown below in case you're not there yet.

$$\mathbf{a} \times \mathbf{b} = (2\mathbf{i} + 5\mathbf{j} + 10\mathbf{k}) \times (7\mathbf{i} - 5\mathbf{j} - 3\mathbf{k})$$

$$= -10\mathbf{k} - 6(-\mathbf{j}) + 35(-\mathbf{k}) - 15\mathbf{i} + 70\mathbf{j} - 50(-\mathbf{i})$$

$$= 35\mathbf{i} + 76\mathbf{j} - 45\mathbf{k}$$

Here's the step - by - step in case you're having
trouble picking it up:

$$2\mathbf{i} \times 7\mathbf{i} = 0 \qquad\qquad 5\mathbf{j} \times 7\mathbf{i} = 35(\mathbf{-k}) \qquad\qquad 10\mathbf{k} \times 7\mathbf{i} = 70\mathbf{j}$$

$$2\mathbf{i} \times (\text{-}5\mathbf{j}) = -10\mathbf{k} \qquad\qquad 5\mathbf{j} \times (-5\mathbf{j}) = 0 \qquad\qquad 10\mathbf{k} \times (-5\mathbf{j}) = -50(-\mathbf{i})$$

$$2\mathbf{i} \times (\text{-}3\mathbf{k}) = \text{-}6(\text{-}\mathbf{j}) \qquad\qquad 5\mathbf{j} \times (-3\mathbf{k}) = -15\mathbf{i} \qquad\qquad 10\mathbf{k} \times (-3\mathbf{k}) = 0$$

3. $\mathbf{a} = \mathbf{i} + 3\mathbf{k}$, $\mathbf{b} = \text{-}5\mathbf{j} - 3\mathbf{k}$

This problem is considerably easier--because there are fewer terms. Generally speaking, the cross product is the hardest of the vector operations. The best thing you can do is get use to using the above diagram by running through a bunch of problems with it.

$$\mathbf{a} \times \mathbf{b} = (\mathbf{i} + 3\mathbf{k}) \times (\text{-}5\mathbf{j} - 3\mathbf{k})$$

$$= -5\mathbf{k} - 3(\text{-}\mathbf{j}) - 15(-\mathbf{i})$$

$$= 15\mathbf{i} + 3\mathbf{j} - 5\mathbf{k}$$

Additional Problems
11.3

Add the given vectors.

1. $\mathbf{a} = 6\mathbf{i} + 2\mathbf{j}$, $\mathbf{b} = 12\mathbf{i} + 7\mathbf{j}$

3. $\mathbf{a} = 12\mathbf{i} - 4\mathbf{j}$, $\mathbf{b} = 2\mathbf{i} - \mathbf{j} + 9\mathbf{k}$

2. $\mathbf{a} = 15\mathbf{j}$, $\mathbf{b} = \mathbf{i}$

4. $\mathbf{a} = \text{-}23\mathbf{i} - 6\mathbf{j} + 3\mathbf{k}$, $\mathbf{b} = 12\mathbf{i} + 5\mathbf{j} + 9\mathbf{k}$

Subtract the given vectors.

5. $\mathbf{a} = 16\mathbf{i} + 7\mathbf{j} + 2\mathbf{k}$, $\mathbf{b} = 8\mathbf{i} - 2\mathbf{j} + \mathbf{k}$

6. $\mathbf{a} = \text{-}\mathbf{i} + \mathbf{k}$, $\mathbf{b} = 13\mathbf{i} - 7\mathbf{j} - 23\mathbf{k}$

Find the dot and cross products of the following vectors.

7. $\mathbf{a} = 4\mathbf{i} + 9\mathbf{j}$, $\mathbf{b} = 12\mathbf{i} + 4\mathbf{j}$

9. $\mathbf{a} = 8\mathbf{k}$, $\mathbf{b} = \mathbf{i} - \mathbf{j}$

8. $\mathbf{a} = \text{-}3\mathbf{i} - 2\mathbf{j}$, $\mathbf{b} = 12\mathbf{i} + 4\mathbf{j} + 3\mathbf{k}$

10. $\mathbf{a} = 20\mathbf{i} - 20\mathbf{j} + 50\mathbf{k}$, $\mathbf{b} = 13\mathbf{i} + 6\mathbf{j} - 12\mathbf{k}$

Vector Functions

In the last chapter we learned about a new type of mathematical variable known as the vector. In this chapter we go a step farther and learn about vector functions (functions that are made-up of vectors), and apply some calculus to them. Since vector functions are so commonly seen in the sciences, we get the added bonus having plenty of real world applications to work with as well.

We began the last chapter by telling you we were going to take a break from calculus. Well, the break is over. While we won't be seeing any new calculus in this chapter, we will be applying some of the calculus we've already learned to functions composed of vectors (the new type of mathematical variables we saw in the last chapter). Fortunately for us, once we know the basics of vector functions, applying calculus to them is easy. You see, there's really not that much to the actual calculations-- nothing extraordinarily difficult like chapters 4, 6, or 8. The most difficult thing here is getting an understanding of the new terms, notation, and environment. Once you've got that, everything falls into place.

12.1 Vector Functions--The Basic Idea

Let's cut right to the chase. As we alluded to in the introduction, vector functions are functions that are made-up of vectors. Specifically, *we define vector functions as functions whose domain is composed of real numbers and whose range is composed of vectors.* Mathematically, we represent vector functions as follows:

$$\mathbf{r}(t) = f(t)\mathbf{i} + g(t)\mathbf{j} + h(t)\mathbf{k}$$

Admittedly, at first glance this doesn't look good. In fact, this doesn't look much like a function. If you think back to the definition of a function, though, you'll see that it is. Here we go:

Recall from Chapter 1 that a function is defined as a mathematical rule that tells us how two groups of numbers are related to each other. Numbers from one group, the domain, are input to the function, which modifies them according to its rule, and then outputs the results into the other group, the range. See Fig. 12.1.

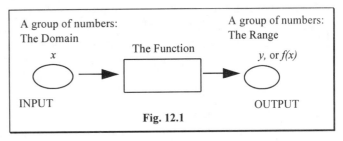

Up until now, the mathematical rule was very simple, example $f(x) = 2x + 1$. In vector functions, however, the rule is not so simple. Numbers are input to the function from the domain as usual, but the function's output is a vector. You see, when numbers from the domain are input to the vector function, they are run through several different functions, the outputs of which make up the components of resulting vectors in the range. See Fig. 12.2.

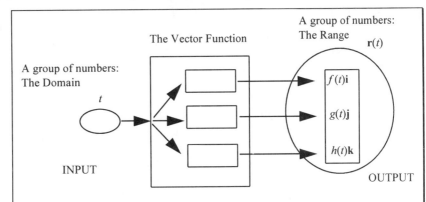

Fig. 12.2: Note: We have single numbers input from the domain into the vector function, which is made up of multiple functions, in this case 3. Each function operates on the input, then outputs the result to the range where they are grouped together to form a vector.

If we look at an example of a vector function completely written out, it starts to make a little more sense:

$$\mathbf{r}(t) = (t^2 + 1)\mathbf{i} + \left(\frac{8-t}{t^3}\right)\mathbf{j} + (7t)\mathbf{k}$$

--where, $f(t) = (t^2 + 1)$, $g(t) = \left(\frac{8-t}{t^3}\right)$, and $h(t) = (7t)$

This looks a lot more reasonable, yes? The domain of the above vector function is composed of the real numbers (except zero) which are represented by the variable t, and the range is made up of vectors. The fact that there are three functions making up the vector function in the above case is only a conceptual difficulty--not a computational one.

Bonus Info: Anytime you have two or more functions with the same independent variable, the independent variable is referred to as a *parameter* and the functions as *parametric equations.* In the above example, *t* is thus a parameter and the three functions, parametric equations. You will often hear these terms in laboratory situations.

On occasion vector functions will be written a bit differently than above--we'll see them in matrix notion. For example, the above vector function can be written:

$$r(t) = \begin{bmatrix} f(t) \\ g(t) \\ h(t) \end{bmatrix} = \begin{bmatrix} t^2 + 1 \\ \dfrac{8-t}{t^3} \\ 7t \end{bmatrix}$$

Note that in matrix notation we don't write out i, j, or k because they are implied by the positions of their corresponding magnitudes. By convention, i is always on top, j in the middle, and k on the bottom.

Or, equivalently, they can be written as:

$$r(t) = \left(f(t),\ g(t),\ h(t) \right) = \left(t^2 + 1,\ \frac{8-t}{t^3},\ 7t \right)$$

Again, we don't write i, j, or k because they are implied by the positions of their corresponding magnitudes. By convention, i is to the left, j in the middle, and k to the right.

Here's an example showing the use of both the regular and matrix notations to solve a problem involving the manipulation of vector functions:

Given the following:

$$s(t) = (t^2 + 1)\mathbf{i} + (9t)\mathbf{j} + (2t)\mathbf{k}$$
$$\mathbf{a} = 2\mathbf{i} - 3\mathbf{j} + 5\mathbf{k}$$

Find the vector, $r(t) = 5s(t) + \mathbf{a}$

Solution:

$$r(t) = 5\left[(t^2 + 1)\mathbf{i} + (9t)\mathbf{j} + (2t)\mathbf{k} \right] + \left[2\mathbf{i} - 3\mathbf{j} + 5\mathbf{k} \right] = (5t^2 + 7)\mathbf{i} + (45t - 3)\mathbf{j} + (10t + 5)\mathbf{k}$$

Alternatively, you could solve as:

$$r(t) = 5\begin{bmatrix} t^2 + 1 \\ 9t \\ 2t \end{bmatrix} + \begin{bmatrix} 2 \\ -3 \\ 5 \end{bmatrix} = \begin{bmatrix} 5t^2 + 5 \\ 45t \\ 10t \end{bmatrix} + \begin{bmatrix} 2 \\ -3 \\ 5 \end{bmatrix} = \begin{bmatrix} 5t^2 + 7 \\ 45t - 3 \\ 10t + 5 \end{bmatrix}$$

Somemore terms: *Vector space.* You're going to hear this term often if you proceed on in the sciences. While the term sounds impressive, it's just another name for dimensions.

For example, a vector in "two space", or "2-vector space", or some combination of terms like that simply means a two dimensional vector--e.g., **i** and **j** only. A vector in "three space", or "3-vector space" simply means a three dimensional vector--e.g., **i**, **j**, and **k**. When you're told to draw a vector in 2 space, don't freak out. Just throw up your garden variety Cartesian coordinate system, draw your vector, and you're golden.

12.2 Taking Limits, Derivatives, and Integrals of Vector Functions

Now we apply some calculus to these vector functions we've been learning about. This is not hard. We begin with the rules.

Taking Limit of Vector Functions:

1. Simply take the limit of each component of the vector:

$$\lim_{t \to a} \mathbf{s}(t) = \left[\lim_{t \to a} f(t) \right] \mathbf{i} + \left[\lim_{t \to a} g(t) \right] \mathbf{j} + \left[\lim_{t \to a} h(t) \right] \mathbf{k}$$

Taking Derivatives of Vector Functions:

1. Simply take the derivative of each component of the vector:

$$\mathbf{s}'(t) = f'(t)\mathbf{i} + g'(t)\mathbf{j} + h'(t)\mathbf{k}$$

2. Some theorems to know when given two vectors, say **s** and **r**:

$$D_t \left[\mathbf{s}(t) + \mathbf{r}(t) \right] = \mathbf{s}'(t) + \mathbf{r}'(t)$$
$$D_t \left[\mathbf{s}(t) \cdot \mathbf{r}(t) \right] = \mathbf{s}'(t) \cdot \mathbf{r}(t) + \mathbf{r}'(t) \cdot \mathbf{s}(t)$$
$$D_t \left[\mathbf{s}(t) \times \mathbf{r}(t) \right] = \mathbf{s}'(t) \times \mathbf{r}(t) + \mathbf{r}'(t) \times \mathbf{s}(t)$$

Taking Integrals of Vector Functions:

$$\int_a^b \mathbf{s}(t)dt = \int_a^b f(t)dt \ \mathbf{i} + \int_a^b g(t)dt \ \mathbf{j} + \int_a^b h(t)dt \ \mathbf{k}$$

As you can see from the rule boxes, all we're really doing in each case is applying the given mathematical operations to each component of the vector. The rules regarding derivatives might require some review, but other than that it's really basic.

Note: Generally speaking, in first year calculus you're not going to be asked to compute limits here, and only rarely integrals. The main focus is on derivatives--because that's where the practical applications come in. Now, the practical applications (and the mathematical ones) that you'll run into will vary tremendously from program to program. If you understand the fundamentals presented here, however, you should be OK as every application is based on them.

Examples   12.2

1. Given $s(t) = 8t^3\mathbf{i} + e^t\mathbf{j} + \sin t\mathbf{k}$, find $s'(t)$ and $s''(t)$.

   Just like the rule says, we take the derivative of each of the vector's components. Not difficult. You won't run into anything here you haven't already seen.

   $s'(t) = 24t^2\mathbf{i} + e^t\mathbf{j} + \cos t\mathbf{k}$

   $s''(t) = 48t\ \mathbf{i} + e^t\mathbf{j} - \sin t\mathbf{k}$

2. Find the integral of $s(t) = 8t^3\mathbf{i} + t^5\mathbf{j} + \sin t\ \mathbf{k}$, from 0 to 1.

   Basically the same thing here, just this time we're using an integral instead of a derivative. The key point to remember is we integrate each component individually, then add the results.

   $$\int_0^1 s(t)dt = \int_0^1 8t^3\mathbf{i} + \int_0^1 t^5\mathbf{j} + \int_0^1 \sin t\ \mathbf{k}$$

   $$= 2t^4\Big|_0^1 + \frac{1}{6}t^6\Big|_0^1 - \cos t\Big|_0^1$$

   $$= 2 - 0 + \frac{1}{6} - 0 - (0.999 - 1) = 2.167$$

Additional Problems   12.2

1. Find $s'(t)$ and $s''(t)$ given, $s(t) = 8t^2\mathbf{i} - 8t^8\mathbf{j}$;

2. Find $s'(t)$ and $s''(t)$ given, $s(t) = e^t\mathbf{i} - \left[18 + (t-4)^2\right]\mathbf{j}$

3. Find the integral of, $s(t) = 8t^2\mathbf{i} - 8t^8\mathbf{j}$, from -1 to 1.

4. Find the integral of, $s(t) = e^t\mathbf{i} - \left[18 + (t-4)^2\right]\mathbf{j}$, from 0 to 5.

12.3 Applications: Velocity and Acceleration

In many real world applications the path a particular object travels through in space can be expressed as a vector function whose domain is time and whose output vectors give the coordinates of the object at any point in time. What we often need to do, given this information, is determine the object's velocity and acceleration at particular points in time--which we can do using derivatives. Here's a typical example:

Suppose your friend Joe picks up a baseball and heaves it 100 yards such that it follows the path specified by the vector function,

$$\mathbf{s}(t) = 10t\,\mathbf{i} + \left[25 - (t-5)^2\right]\mathbf{j};$$

which is shown graphically in Fig. 12.3.

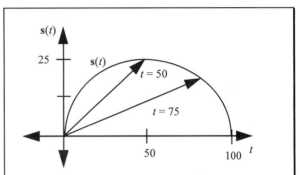

Fig. 12.3: Graph of the vector function, $\mathbf{s}(t)$, which gives the position (coordinates) of Jim's ball, at any time. Note we've only explicitly drawn out two of the vectors (the ones at $t = 50$ and $t = 75$), because it would be too congested otherwise. The curve, which shows the path the ball took, comes from connecting the heads of all the vectors.

If we want to determine the velocity of the ball at any particular point during its flight, the first thing we have to do is take the derivative of the above function. Recall from Chapter 4 that when we take the derivative of a function representing distance it gives us speed. Here, however, when we take the derivative we get velocity because both magnitude and direction are specified in the vector (Remember, velocity is not only a measure of an object's speed, its also a measure of the direction the object is moving). *In general then, when we take the derivative of a vector function which represents position, we get another vector function which represents velocity.*

So, when we take the derivative of the above function, we get:

$$\mathbf{s}'(t) = \mathbf{v}(t) = 10\,\mathbf{i} - 2(t-5)\,\mathbf{j}$$

which represents, as a vector, the velocity of the ball at every point during its flight.

Note: Velocity vectors are tangent to their corresponding points on the position graph. Why? Because the derivative of a function tells you the slopes of the lines tangent to the points of the function. Fig. 12.4 shows a few of the velocity vectors for our above function, at $t = 1$, $t = 5$, and $t = 8$.

As you will recall from Chapter 4, if we wish to continue and determine the acceleration of the ball at any particular point during its flight, we need to take the second derivative of our position function. Doing this, we obtain:

$$\mathbf{s''}(t) = \mathbf{a}(t) = -2\,\mathbf{j}$$

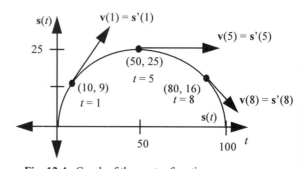

Fig. 12.4: Graph of the vector function, s(t), with velocity vectors at t = 1, t = 5, and t = 8 shown.

Hence, we see in this case that acceleration is constant in the negative **j** direction throughout the ball's flight. (This makes sense because gravity constantly acts downward!)

Occasionally we'll want to see all three quantities--position, velocity, and acceleration--on the same graph. To do this, we typically draw out the entire position graph, then pick out particular points of interest and draw the velocity and acceleration vectors at those points. This is not particularly difficult as long as you're graph is drawn pretty close to scale.

For example, if we wanted to see graphically what was going on at t = 3 in terms of position, velocity and acceleration of the ball Jim heaved, we'd resketch the position graph, then draw in the velocity and acceleration vectors at t = 3. To do this we simply solve **s'**(3) and **s''**(3) for the vectors, then plot. Fig. 12.5 shows the results.

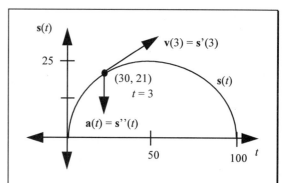

Fig. 12.5: Graph of the vector function, s(t), showing the velocity and acceleration vectors at t = 3.

Real World Application: Suppose you and you're friend have designed a new golf ball which you think will travel farther than any other golf ball out in the market. In order to test your belief, you videotape your's and competitors' balls being hit with a machine that delivers an identical impact force with every blow. After videotaping various balls throughout their flights, you run the data through a computer via the aid of a motion analysis video processing machine and come up with vector functions representing the paths of each ball. As part of your analysis, undoubtedly some of the information you'd want to look at would be the velocity and acceleration at various points in time during the balls' flights.

Real World Application: Suppose you're employed for a private company that sends satellites into orbit. Your boss comes up to you and asks you to determine if a given orbit for a particular satellite will be adequate, or if a higher orbit is necessary. In making your determination, clearly one of the things you'd have to do is calculate the velocity and acceleration of the satellite at various points in the proposed orbit. Since orbits can be expressed as vector functions, this is no problem.

Examples   12.3

Find the speed, velocity, and acceleration of the given objects at time t, then sketch their path with velocity and acceleration vectors shown at time t.

1. During the course of the shortened 1994 baseball season there were claims made by sports writers that the baseballs being used were "juiced" due to the fact that many of the players were hitting substantially better than in previous years. Of course, many sports fans, and other writers for that matter, disagreed with this juicing theory, claiming instead that the hitters were getting better and the pitchers getting worse.

Suppose, in an effort to resolve this controversy, you obtain a bag of baseballs from 1993 and a bag of baseballs from 1994, then shoot them out of a pitching machine that's aimed high. You videotape the baseballs as they fly through the air--just as you did with the golf balls--then analyze the path data for each ball via computer and come up with a vector function for each. Given the following vector function representing the path of ball A, solve for its speed, velocity, and acceleration at time $t = 3$, then plot as specified above.

Given: $\mathbf{s}(t) = 25t\,\mathbf{i} + \left[16 - (t-4)^2\right]\mathbf{j}$, then

speed:

$$|\mathbf{s'}(3)| = \sqrt{(25)^2 + (2)^2} = 25.0$$

velocity:

$\mathbf{s'}(t) = 25\,\mathbf{i} - 2(t-4)\,\mathbf{j}$

$\mathbf{s'}(3) = 25\,\mathbf{i} - 2(3-4)\,\mathbf{j} = 25\,\mathbf{i} + 2\,\mathbf{j}$

acceleration:

$\mathbf{s''}(t) = -2\,\mathbf{j}$

$\mathbf{s''}(3) = -2\,\mathbf{j}$

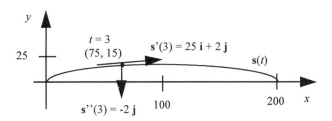

Plotting is not a trivial thing in regards to the path. The fastest way to do it is brute force--i.e., plug in numbers and plot. Once that's done, you just roughly draw in the vectors (don't worry about them being to scale, just make sure you label everything).

Additional Problems 12.3

Find the speed, velocity, and acceleration of the given objects at time t, then sketch their path with velocity and acceleration vectors shown at time t.

1. $s(t) = \dfrac{1}{t}\,\mathbf{i} + \left[\dfrac{2}{t+2}\right]\mathbf{j}$, $t = 3$

2. $s(t) = 3t\,\mathbf{i} + e^t\,\mathbf{j}$, $t = 5$

12.4 Arc Length and Curvature

Now that we're familiar with the fact that we can express the path of an object as a vector function, it's time to learn how to calculate the length of the path. What do we mean by this? Well, suppose we have a piece of string. If we pull the string straight, we have no problem measuring its length, right? But what if we lay the string down on the ground and contort it? While the length of the string hasn't changed, calculating the length now is considerably more difficult. At first glance then, our task of calculating the lengths of paths, like the one in Fig. 12.6 that are not straight, seems pretty complicated. Fortunately, if the path is given as a vector function, all we have to do is use the following formula:

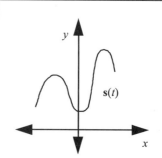

Fig. 12.6: The path of an object with tangent and normal vectors shown.

Given that the vector function, $s(t) = f(t)\mathbf{i} + g(t)\mathbf{j} + h(t)\mathbf{k}$, represents the path of an object, the length of the path is:

$$L = \int_a^b \left(\sqrt{[f(t)]^2 + [g(t)]^2 + [h(t)]^2} \right) dt$$

Another bit of information that we sometimes need to get about the path of an object has to do with unit vectors. On occasion, it's useful to know the unit vectors that are tangent and normal to the points that make up the path. What do we mean by "normal"? We mean perpendicular to the tangent, in the direction towards the center of curvature. See Fig. 12.7. Why would we want to know about these unit vectors? Because in physics, when we're analyzing an object's motion in great detail, it's often desirable to analyze the velocity and acceleration of the object in terms of its tangential and normal components.

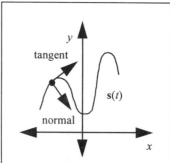

Fig. 12.7: The path of an object with tangent and normal vectors shown.

So how do we go about obtaining the unit vectors that are tangential and normal to the points that make up the path of an object? Simply follow these rules:

Unit vector tangent to path:

If $s(t)$ is a vector function representing the path of an object, then the unit vector tangent to the path is given by,

$$T(t) = \frac{1}{|s'(t)|} s'(t)$$

Unit vector normal to path:

If $s(t)$ is a vector function representing the path of an object and $T(t)$ the unit vector tangent to the path, then the unit vector normal to the path is given by,

$$N(t) = \frac{1}{|T'(t)|} T'(t)$$

Examples   12.4

Calculate the lengths of the given paths.

1. $s(t) = 8t\,\mathbf{i} + 9t\mathbf{j}$, 0 to 2

To solve for the length of the path, simply follow the above formula--this is just plug'n chug.

$$L = \int_0^2 \left(\sqrt{(8t)^2 + (9t)^2} \right) dt$$

$$= \int_0^2 \left(\sqrt{145t^2} \right) dt$$

$$= \int_0^2 \left(\sqrt{145t^2} \right) dt$$

$$= \int_0^2 \left(t\sqrt{145} \right) dt = \frac{\sqrt{145}}{2} t^2 \Big|_0^2 = \left(\frac{\sqrt{145}}{2}(2)^2 - 0 \right) = 2\sqrt{145}$$

Calculate the tangential and normal unit vectors for the given vector functions.

2. $s(t) = 8t\,\mathbf{i} + 2t^2\,\mathbf{j}$

$$T(t) = \frac{1}{\sqrt{(8)^2 + (4t)^2}}(8\,\mathbf{i} + 4t\,\mathbf{j})$$

$$= \frac{1}{\sqrt{64 + 16t^2}}(8\,\mathbf{i} + 4t\,\mathbf{j})$$

$$= \frac{1}{4\sqrt{4 + t^2}}(8\,\mathbf{i} + 4t\,\mathbf{j})$$

$$= \frac{2}{\sqrt{4 + t^2}}\mathbf{i} + \frac{1}{\sqrt{4 + t^2}}\mathbf{j}$$

Note: When solving for the tangential and normal unit vectors things get hairy in a hurry. While the formulas look innocent enough, their output is heinous. Usually the tangential unit vector can be calculated with relative ease, as we show here, just following the formula. However, as you can see from our work below, calculating the normal unit vector is a job. As long as you follow the formulas though, you can do it-- it's just a lot of baggage to work with.

$$N(t) = \frac{1}{\sqrt{\left(\dfrac{-2\left(\frac{1}{2}\right)(4+t^2)^{-\frac{1}{2}}(2t)}{\left(\sqrt{4+t^2}\right)^2} + \dfrac{-1\left(\frac{1}{2}\right)(4+t^2)^{-\frac{1}{2}}(2t)}{\left(\sqrt{4+t^2}\right)^2}\right)}} \cdot$$

$$\left(\frac{-2\left(\frac{1}{2}\right)(4+t^2)^{-\frac{1}{2}}(2t)}{\left(\sqrt{4+t^2}\right)^2}\mathbf{i} + \frac{-1\left(\frac{1}{2}\right)(4+t^2)^{-\frac{1}{2}}(2t)}{\left(\sqrt{4+t^2}\right)^2}\mathbf{j}\right)$$

$$= \frac{1}{\sqrt{\left(\dfrac{-2t}{(4+t^2)\sqrt{4+t^2}} - \dfrac{t}{(4+t^2)\sqrt{4+t^2}}\right)}}\left(\frac{-2t}{(4+t^2)\sqrt{4+t^2}}\mathbf{i} - \frac{t}{(4+t^2)\sqrt{4+t^2}}\mathbf{j}\right)$$

$$= \frac{1}{\sqrt{\dfrac{-2t}{(4+t^2)\sqrt{4+t^2}}}}\left(\frac{-2t}{(4+t^2)\sqrt{4+t^2}}\mathbf{i} - \frac{t}{(4+t^2)\sqrt{4+t^2}}\mathbf{j}\right)$$

Additional Problems   12.4

Calculate the lengths of the following vector functions, as well as their tangential and normal unit vectors.

1. $s(t) = t\,\mathbf{i} + t^2\,\mathbf{j}$

2. $s(t) = 36t^3\,\mathbf{i} + 2t^3\,\mathbf{j}$

Partial Derivatives & Multiple Integrals

In this chapter we get a sneak preview of third semester math--calculus applied to multivariable functions. In this chapter we concentrate on how to take derivatives of functions containing more than one independent variable, and of integrals of functions with two independent variables. Enjoy the sneak preview, first year calculus is over. Rejoice!

F olks, we are standing on the brink of complete success. All that is left for us to do in this first year of calculus is take a look at what lies ahead in third semester math. You won't find anything difficult in this chapter. All we're doing is expanding our discussion of calculus to multivariable functions. Since in many real life situations the relevant functions depend on more than one independent variable, it's rather important that we know how to apply the calculus we've learned in this first year to them. In third semester you will get into this topic in much more detail. As long as you remember that it's nothing more than the calculus you've already learned applied to multivariable functions, though, you should have no trouble.

13.1 Partial Derivatives--The Basic Idea

Throughout this first year of calculus we've learned to apply the three mathematical operations that comprise calculus--limits, derivatives, and integrals--to functions with only one independent variable. For example, functions like:

$$y = f(x) = 2x^2 + 6$$

In third semester calculus this changes. Instead of applying calculus to single variable functions, we apply it to multivariable functions--i.e., functions that depend on more than one independent variable, like:

$$y = f(x,z) = 3x^3 + xz^2 + 3x - 7z + 3$$

As you might expect, doing this results in some not so pleasant complications. We begin our discussion with the effects on derivatives.

Taking the derivatives of single variable functions is, as we saw in Chapter 3, not that hard. Unfortunately, when we take the derivatives of multivariable functions things get a bit more complicated and we have to be careful. Why? Because there's more than one possible derivative to take. In fact, there's as many possible derivatives as there are independent variables. Hence, we have to make sure we know what we're looking for at the onset of the problem.

Example: Suppose we have a metal wire that we heat at one end. Obviously, the end of the wire closer to the flame is hotter than the other end. We can describe the temperature at any point along the wire mathematically via the function, $y = f(x)$, where y is the temperature and x is the location of the point (See Fig. 13.1). Hence, if we take the derivative of this function, we obtain an expression for the rate at which the temperature is changing per unit of distance along the wire--e.g., In taking the derivative we may find that the temperature decreases at a rate of two degrees per inch as we move along the wire away from the flame.

Fig. 13.1: A wire being heated at one end. The variable x completely describes the points along the wire since the wire is only 1 dimension.

Suppose we change the above scenario by keeping track of time as well. In other words, now we have a function, $y = f(x, t)$, where x is the location of the point and t is the time starting from when the heat was first applied to the wire. If we want to take the derivative of this function, we first have to know what we're looking for: The rate at which the temperature is changing per unit of distance along the wire (derivative #1), or the rate at which the temperature is changing at a particular point over the course of time (derivative #2). You see, when we have more than one independent variable in a function, we can't just "take the derivative" of the function. We first have to decide which one. Not surprisingly, the derivatives of functions with more than one independent variable have special names--they're referred to as *partial derivatives*.

How do we take partial derivatives? We simply determine which variable we're interested in finding the derivative of and then take the derivative of the function with respect to it while holding the other variable constant. So, in our above example, if we wanted to determine the rate at which the temperature was changing per unit of distance along the wire, we would take the derivative of the function with respect to x and treat t as a constant. Similarly, if we wanted to determine the rate at which the temperature was changing at a particular point, we would take the derivative of the function with respect to t and treat x as a constant.

Notation:

If $y = f(x,z)$, then:

1. The partial derivative of f with respect to x is, $f_x(x,z) = \dfrac{\partial y}{\partial x}$

2. The partial derivative of f with respect to z is, $f_z(x,z) = \dfrac{\partial y}{\partial z}$

Examples   **13.1**

Find the first partial derivatives of the given functions.

1. $y = f(x,z) = 9x^2 + 13xz^2 - 9z + 5$

$f_x(x,z) = \dfrac{\partial y}{\partial x} = 18x + 13z^2 - 0 + 0$

$f_z(x,z) = \dfrac{\partial y}{\partial z} = 0 + 26xz - 9 + 0$

This is not hard. All you do is decide which variable's derivative you're going to find, then hold the other variable constant and go.

Here, we solve for the partial derivative of x first, hence z is treated as if it were a constant. Once that's done, we solve for the partial derivative of z and treat x as if it were a constant.

2. $y = f(x,z) = e^x \ln xz$

$f_x(x,z) = \dfrac{\partial y}{\partial x} = e^x \ln xz + e^x \left(\dfrac{1}{xz}\right) z = e^x \ln xz + \dfrac{e^x}{x}$

$f_z(x,z) = \dfrac{\partial y}{\partial z} = e^x \left(\dfrac{1}{xz}\right) x = \dfrac{e^x}{z}$

Same thing as above. Note that all the rules for finding derivatives apply here just as they did throughout first year calculus. The only difference is in how we handle the multiple variables.

3. $y = f(x,z,r) = 3x^3 + 3xz^2 r^3 + rx^2 + 3x \cos z$

$f_x(x,z,r) = \dfrac{\partial y}{\partial x} = 9x^2 + 3z^2 r^3 + 2rx + 3\cos z$

$f_z(x,z,r) = \dfrac{\partial y}{\partial z} = 0 + 6xzr^3 + 0 - 3x \sin z$

$f_r(x,z,r) = \dfrac{\partial y}{rz} = 0 + 9xz^2 r^2 + x^2 + 0$

Slightly more complex problem due to the fact that we have 3 independent variables now instead of just 2. No change in the basic solving pattern, however, just more to do. The basic rule stays the same--decide which variable you're finding the derivative of, then treat the others as if they were constants.

Second Partial Derivatives

As with normal derivatives, we can take higher order partial derivatives as well. We will limit our discussion to just second order partial derivatives since that will adequately illustrate our point here.

Finding second order partial derivatives is easy if you just remember that each first order partial derivative has as many second order partial derivatives as it has independent variables. For example, a function $y = f(x, z)$ has the following first and second order partial derivatives:

1st order partial derivatives of $f(x,z)$:

$f_x(x,z)$

$f_z(x,z)$

2nd order partial derivatives of $f(x,z)$:

$f_{xx}(x,z)$, which is the partial derivative of $f_x(x,z)$ taken with repect x.

$f_{xz}(x,z)$, which is the partial derivative of $f_x(x,z)$ taken with repect z.

$f_{zz}(x,z)$, which is the partial derivative of $f_z(x,z)$ taken with repect z.

$f_{zx}(x,z)$, which is the partial derivative of $f_z(x,z)$ taken with repect x.

Note: $f_{xz}(x,z) = f_{zx}(x,z)$.

Formal Notation:

$$f_{xx}(x,z) = \frac{\partial}{\partial x}\left(\frac{\partial y}{\partial x}\right) = \frac{\partial^2 y}{\partial x^2}$$

$$f_{xz}(x,z) = \frac{\partial}{\partial x}\left(\frac{\partial y}{\partial z}\right) = \frac{\partial^2 y}{\partial x \partial z}$$

$$f_{zz}(x,z) = \frac{\partial}{\partial z}\left(\frac{\partial y}{\partial z}\right) = \frac{\partial^2 y}{\partial z^2}$$

$$f_{zx}(x,z) = \frac{\partial}{\partial z}\left(\frac{\partial y}{\partial x}\right) = \frac{\partial^2 y}{\partial z \partial x}$$

Admittedly, this formal notation looks nasty, but don't be fooled. The actual mechanics of the problem solving is easy. Once you get use to the squiggly d's, it's not that bad. Concentrate on the examples, get them down, and the notation will come.

Examples 13.1.2

Find the second partial derivatives of the given function.

1. $y = f(x,z) = 9xz^2 + x\cos z + 3z + x^2$

This looks like a lot, but it's really not. Just follow the rules on the opposite page, and make real sure you get the first partials correct or all your second partials will be messed up.

$$f_x(x,z) = \frac{\partial y}{\partial x} = 9z^2 + \cos z + 2x$$

$$f_z(x,z) = \frac{\partial y}{\partial z} = 18xz - x\sin z + 3$$

The key, again, is to decide what variable's derivative you're taking, then treat the others as constants.

$$f_{xx}(x,z) = \frac{\partial^2 y}{\partial x^2} = 2$$

$$f_{xz}(x,z) = \frac{\partial^2 y}{\partial z \partial x} = 18z - \sin z$$

$$f_{zz}(x,z) = \frac{\partial^2 y}{\partial z^2} = 18z - x\cos z$$

$$f_{zx}(x,z) = \frac{\partial^2 y}{\partial x \partial z} = 18z - \sin z$$

Additional Problems 13.1

Find the first and second partial derivatives of the following functions.

1. $y = f(x,z) = 2z + 18x^2z^3 + 2e^xz$ 2. $y = f(x,z) = 2(x + 2z)^3$

13.2 The Chain Rule for Partial Derivatives

Now suppose we have a function that depends on other functions (in other words, a function of a function), not just variables. For example, say we're given a function, $g = f(r, p)$ where $r = h(x, z)$ and $p = k(x, z)$. How do we go about finding the partial derivatives of g with respect to x and z? Simple, we just follow the chain rule:

$$f_x(r,p) = \frac{\partial g}{\partial x} = \frac{\partial g}{\partial r} \cdot \frac{\partial r}{\partial x} + \frac{\partial g}{\partial p} \cdot \frac{\partial p}{\partial x}$$

$$f_z(r,p) = \frac{\partial g}{\partial x} = \frac{\partial g}{\partial r} \cdot \frac{\partial r}{\partial z} + \frac{\partial g}{\partial p} \cdot \frac{\partial p}{\partial z}$$

Essentially what we're doing here is taking the specified partial derivatives and then adding or multiplying them together. The reason we have to do this is because we can't get to x or z directly.

Again we have a situation with notation that makes the situation look ugly. Note, however, that all we're doing is working our way through the "layers" to get to the variables. It's really nothing more than a little additional multiplying and addition. Check out the examples--it's the best way to learn this.

Examples   13.2
Given the function g, find its partial derivatives.

1. $g = f(r,p) = r^2 + p^2, \quad r = x^2 + 2xz, \quad p = 3z + x$

The trick here is to find each partial derivative as specified by the chain rule formula, then just plug them in, multiply and add.

$$f_x(r,p) = \frac{\partial g}{\partial x} = \left(\frac{\partial g}{\partial r}\right)\left(\frac{\partial r}{\partial x}\right) + \left(\frac{\partial g}{\partial p}\right)\left(\frac{\partial p}{\partial x}\right)$$

$$= (2r)(2x + 2z) + (2p)(1)$$

$$= (2x^2 + 4xz)(2x + 2z) + (6z + 2x)$$

Here, we employ the chain rule following the pattern shown above, solve for the specified partials, then just plug them in and make the appropriate substitutions to get rid of the p's and r's.

$$f_z(r,p) = \frac{\partial g}{\partial z} = \left(\frac{\partial g}{\partial r}\right)\left(\frac{\partial r}{\partial z}\right) + \left(\frac{\partial g}{\partial p}\right)\left(\frac{\partial p}{\partial z}\right)$$

$$= (2r)(2x) + (2p)(3)$$

$$= (2x^2 + 4xz)(2x) + (6z + 2x)(3)$$

2. $g = f(r,p,q) = rp^2 + q^2 r^3 - 19q, \quad r = x^2 + z^2, \quad p = 3x - z, \quad q = xz$

$$f_x(r,p,q) = \frac{\partial g}{\partial x} = \left(\frac{\partial g}{\partial r}\right)\left(\frac{\partial r}{\partial x}\right) + \left(\frac{\partial g}{\partial p}\right)\left(\frac{\partial p}{\partial x}\right) + \left(\frac{\partial g}{\partial q}\right)\left(\frac{\partial q}{\partial x}\right)$$

$$= (p^2 + 3q^2 r^2)(2x) + (2rp)(3) + (2qr^3 - 19)(z)$$

$$= \left([3x - z]^2 + 3[xz]^2 [x^2 + z^2]^2\right)(2x) + \left(2[x^2 + z^2][3x - z]\right)(3)$$

$$+ \left(2[xz][x^2 + z^2]^3 - 19\right)(z)$$

Note: This problem looks bad, but actually it's really easy. All we're doing here is accounting for another independent variable--following the same pattern as the others no less. The thing that make this problem look bad is all the algebra at the end. Don't get lost in it. Spend most of your time concentrating on the parts where the chain rule is written out-- that's the easiest and the most relevant.

$$f_z(r,p) = \frac{\partial g}{\partial z} = \left(\frac{\partial g}{\partial r}\right)\left(\frac{\partial r}{\partial z}\right) + \left(\frac{\partial g}{\partial p}\right)\left(\frac{\partial p}{\partial z}\right) + \left(\frac{\partial g}{\partial q}\right)\left(\frac{\partial q}{\partial z}\right)$$

$$= (p^2 + 3q^2 r^2)(2z) + (2rp)(-1) + (2qr^3 - 19)(x)$$

$$= \left([3x - z]^2 + 3[xz]^2 [x^2 + z^2]^2\right)(2z) + \left(2[x^2 + z^2][3x - z]\right)(-1)$$

$$+ \left(2[xz][x^2 + z^2]^3 - 19\right)(x)$$

Additional Problems   **13.2**

Given the function, g, find its partial derivatives.

1. $g = f(r,p) = 3r + p^{\frac{1}{2}}$, $r = x^2 + xz$, $p = 3x - z$

2. $g = f(r,p) = 9r^3 + 96p + r^2p^3$, $r = e^x + z$, $p = z^2x - 1$

3. $g = f(r,p,q) = rpq + \ln p + r\sin q$, $r = x + 2$, $p = x^2z$, $q = 1 + z^2$

13.3 The Basis of the Double Integral

Continuing on with this idea of applying calculus to multivariable functions, sometimes we run into a multivariable function and desire to integrate it. In this section we'll introduce a few of the basics behind the approach taken to do this. We'll start with the simplest case:

Suppose we have the function $y = f(x, z)$ and we want to integrate it. The first thing we need to do is determine the bounds of each of the independent variables--in this case x and z. Once we've done this we simply integrate using the same premise as partial derivatives--integrate one variable first while holding the other constant, then integrate the other holding the first constant. It doesn't matter what order we do the integration, so long as the appropriate bounds are used. And note, all of the rules of integration that we've learned apply here as well.

> **Notation:**
> $$\int_a^b \int_c^d f(x,z)\,dxdz = \int_a^b \left[\int_c^d f(x,z)\,dx \right] dz = \int_c^d \left[\int_a^b f(x,z)\,dz \right] dx$$

Examples   **13.3.1**

1. $\int_1^2 \int_{-3}^9 (4x^2z + 5)\,dxdz$

Notice in this problem we begin by integrating x first, treating z as if it were a constant. We ignore the outer integral sign. Just concentrate on the inner one.

$$\int_1^2 \int_{-3}^9 (4x^2z + 5)\,dxdz = \int_1^2 \left[\int_{-3}^9 (4x^2z + 5)\,dx \right] dz$$

$$= \int_1^2 \left[\frac{4}{3}x^3z + 5x \right]_{-3}^9 dz$$

$$= \int_1^2 \left[\left(\frac{4}{3}(9)^3 z + 5(9) \right) - \left(\frac{4}{3}(-3)^3 z + 5(-3) \right) \right] dz$$

$$= \int_1^2 \left[972z + 45 + 36z + 15 \right] dz$$

$$= \int_1^2 (1008z + 60) dz$$

$$= \frac{1008}{2} z^2 + 60z \Big|_1^2 = 504(2)^2 + 60(2) - 504(1)^2 - 60(1) = 1572$$

2. $\int_{-3}^9 \int_1^2 \left(4x^2 z + 5 \right) dz dx$
   Same problem, different order of integration--just to show you it can be done.

$$\int_{-3}^9 \int_1^2 \left(4x^2 z + 5 \right) dz dx = \int_{-3}^9 \left[\int_1^2 \left(4x^2 z + 5 \right) dz \right] dx$$

$$= \int_{-3}^9 \left[2x^2 z^2 + 5z \right] \Big|_1^2 dx$$

$$= \int_{-3}^9 \left[\left(2x^2 (2)^2 z + 5(2) \right) - \left(2x^2 (1)^2 z + 5(1) \right) \right] dz$$

$$= \int_{-3}^9 \left(8x^2 + 10 - 2x^2 - 5 \right) dz$$

$$= \int_{-3}^9 \left(6x^2 + 5 \right) dz$$

$$= 2x^3 + 5x \Big|_{-3}^9 = 2(9)^3 + 5(9) - 2(-3)^3 - 5(-3) = 1572$$

Sometimes the bounds of the integration are functions rather than constants. That's OK, we just proceed as normal--all of the old rules still apply. Using functions as bounds does create on complication, however, when we want to reverse the order of integration. Unlike the normal case of using constants as bounds, when using functions as bounds we can't just switch the order and go. Rather, we have to first come up with new corresponding bounds. Since doing this is a little complicated and is only really applicable when dealing with area and volume problems, we won't deal with these types of problems here. Just be aware that this complication exists so that when you see it in third semester you're not surprised. For now, we'll just concentrate on practicing some straight forward double integrals with functions in their bounds.

Notation:

$$\int_a^b \int_{h_1(x)}^{h_2(x)} f(x,z) dz dx = \int_a^b \left[\int_{h_1(x)}^{h_2(x)} f(x,z) dz \right] dx$$

$$\int_a^b \int_{h_1(z)}^{h_2(z)} f(x,z) dx dz = \int_a^b \left[\int_{h_1(z)}^{h_2(z)} f(x,z) dx \right] dz$$

Examples 13.3

Evaluate the following integrals.

1. $\int_{-1}^{2}\int_{x^2}^{x+2}\left(zx^2-z\right)dzdx$

Note that while having functions as boundaries does make things more messy, it does not change the basic problem solving mechanics. When you're faced with these types of problems, just take a deep breath before you dive in.

$\int_{-1}^{2}\int_{x^2}^{x+2}\left(zx^2-z\right)dzdx=\int_{-1}^{2}\left[\int_{x^2}^{x+2}\left(zx^2-z\right)dz\right]dx$

$\qquad=\int_{-1}^{2}\left[\frac{1}{2}z^2x^2-\frac{1}{2}z^2\right]_{x^2}^{x+2}dx$

$\qquad=\int_{-1}^{2}\left(\left[\frac{1}{2}(x+2)^2x^2-\frac{1}{2}(x+2)^2\right]-\left[\frac{1}{2}(x^2)^2x^2-\frac{1}{2}(x^2)^2\right]\right)dx$

$\qquad=\int_{-1}^{2}\left(\frac{1}{2}(x^4+4x^3+4x^2)-\frac{1}{2}(x^2+4x+4)-\frac{1}{2}x^6+\frac{1}{2}x^4\right)dx$

$\qquad=\int_{-1}^{2}\left(-\frac{1}{2}x^6+x^4+2x^3+\frac{3}{2}x^2-2x-2\right)dx$

$\qquad=\left[-\frac{1}{14}x^7+\frac{1}{5}x^5+\frac{1}{2}x^4+\frac{1}{2}x^3-x^2-2x\right]_{-1}^{2}$

$\qquad=\left[-\frac{1}{14}(2)^7+\frac{1}{5}(2)^5+\frac{1}{2}(2)^4+\frac{1}{2}(2)^3-(2)^2-2(2)\right]$

$\qquad\quad-\left[-\frac{1}{14}(-1)^7+\frac{1}{5}(-1)^5+\frac{1}{2}(-1)^4+\frac{1}{2}(-1)^3-(-1)^2-2(-1)\right]$

$\qquad=0.386$

2. $\int_{0}^{2}\int_{z^2}^{z}\left(xz^2\right)dxdz$

This example is much easier to solve than the previous one simply because the functions in the bounds are more friendly. Remember, though, the point is this--if you just concentrate on the mechanics, everything else will fall into place. Don't let excess algebra confuse you. Once you set the problem up, the rest is trivial number moving, that's all.

$\int_{0}^{2}\int_{z^2}^{z}\left(xz^2\right)dxdz=\int_{0}^{2}\left[\int_{z^2}^{z}\left(xz^2\right)dx\right]dz$

$\qquad=\int_{0}^{2}\left[\frac{1}{2}x^2z^2\right]_{z^2}^{z}dz$

$\qquad=\int_{0}^{2}\left[\frac{1}{2}(z)^2z^2-\frac{1}{2}\left(z^2\right)^2z^2\right]dz$

$\qquad=\int_{0}^{2}\left[\frac{1}{2}z^4-\frac{1}{2}z^3\right]dz$

$\qquad=\left[\frac{1}{10}z^5-\frac{1}{8}z^4\right]_{0}^{2}=1.2$

Additional Problems   **13.3**

Evaluate the following integrals.

1. $\int_{-2}^{0}\int_{3}^{4}\left(5z+4x^2\cos z\right)dxdz$

2. $\int_{3}^{4}\int_{-2}^{0}\left(5z+4x^2\cos z\right)dzdx$

3. $\int_{1}^{2}\int_{z}^{5-z}\left(z+4x\right)dxdz$

4. $\int_{1}^{2}\int_{x^3}^{x}e^{\frac{z}{x}}dxdz$

Solutions to Additional Problems

Chapter 2

Section 2.1
1. 3
2. 5
3. 1020
4. $\dfrac{5}{17}$
5. π
6. 2
7. 6
8. 14
9. -3
10. $\dfrac{1}{2}$
11. 0
12. 11
13. 97
14. 24
15. DNE
16. -6
17. 8
18. 10
19. $\dfrac{1}{9}$
20. $\dfrac{1}{12}$
21. $\dfrac{1}{27}$
22. $\dfrac{1}{75}$
23. $\dfrac{1}{12}$
24. $\dfrac{1}{27}$

25. $\dfrac{19}{\sqrt{3}}$
26. $-\dfrac{3}{8}$
27. $-\dfrac{1}{4}$
28. 8
29. 1
30. 0
31. DNE
32. $-\dfrac{\sqrt{3}}{18}$
33. 0
34. DNE
35. DNE
36. -1
37. $\sqrt[3]{5}$
38. DNE
39. 64
40. must graph
41. -1
42. 0
43. 0
44. $\dfrac{3}{2}$
45. $-\dfrac{1}{8}$
46. $\dfrac{1}{4}$
47. $\dfrac{1}{4}$

Section 2.2
1. 0
2. DNE

3. DNE
4. 14
5. 10
6. 2
7. 0
8. DNE
9. -1
10. 1
11. DNE
12. -1
13. 1
14. DNE
15. 4
16. DNE
17. 3
18. 3
19. 3
20. DNE
21. $\sqrt{6}$
22. 1
23. 1
24. 1

Section 2.3
1. 1
2. 3
3. 1
4. 2
5. 0
6. $\dfrac{1}{6}$
7. 0
8. 1
9. 0

10. 1

11. 0

12. 2

13. 1

14. 1

15. 0

16. $\dfrac{17}{67}$

17. $\dfrac{310}{483}$

18. 0

19. 1

20. 1

21. 0

22. 7

23. 1

Section 2.4

1. continuous
2. continuous
3. discontinuous
4. continuous
5. continuous
6. discontinuous
7. discontinuous
8. continuous
9. continuous
10. discontinuous
11. continuous
12. continuous
13. discontinuous
14. discontinuous
15. discontinuous
16. continuous
17. discontinuous
18. continuous
19. discontinuous
20. continuous
21. $(-\infty,0)(0,\infty)$
22. $(-\infty,1)(1,\infty)$

23. $\left(-\infty,\dfrac{13-\sqrt{161}}{2}\right)$

$\left(\dfrac{13-\sqrt{161}}{2},\dfrac{13+\sqrt{161}}{2}\right)$

$\left(\dfrac{13+\sqrt{161}}{2},\infty\right)$

24. $(-\infty,\infty)$

25. $(-\infty,\infty)$

26. $(-\infty,3)(3,\infty)$

27. $(-\infty,-3)(-3,\infty)$

Chapter 3

Section 3.1

1. 3
2. -7
3. 19
4. -3
5. 0
6. 0
7. $2x+1$
8. $6x-7$
9. $8x-28$
10. $-4x^{-3}$

Section 3.2

1. 0
2. 0
3. 0
4. 0
5. $3x^2$
6. $597x^{596}$
7. $6x$
8. $35x^4$
9. $-16x$
10. $6x+35x^4$
11. $2+21x^2$
12. $-55x^4+3x^2+6x$

13. 1

14. $2x-1$

15. $26x-42x^5+1$

16. $-\dfrac{1}{(x+1)^2}$

17. $\dfrac{14x^3+3x^2+1}{(7x+1)^2}$

18. $\dfrac{-8x^2-2x-3}{x^6}$

19. $-x^{-2}$

20. $-6x^{-7}$

21. $-19x^{-20}$

22. $-6x^{-4}$

23. $21x^{-8}$

24. $-40x^{-3}$

25. $\dfrac{1}{2}x^{-\frac{1}{2}}$

26. $\dfrac{1}{3}x^{-\frac{2}{3}}$

27. $-\dfrac{1}{7}x^{-\frac{8}{7}}$

28. $-\dfrac{1}{2}x^{-\frac{3}{2}}$

29. $-x^{-\frac{4}{3}}$

30. $\cos x$

31. $-10\sin x$
    $+13\cos x$
    $+3x^2+2x^{-3}$

Section 3.3

1. $\Delta y=4\Delta x$
   $dy=5dx$
   $\Delta y-dy=0$

2. $\Delta y = 8x\Delta x + 4\Delta x^2$

   $dy = 8xdx$

   $\Delta y - dy = 4\Delta x^2$

3. $\Delta y = 10x\Delta x - 3\Delta x + 9\Delta x^2$

   $dy = (18x - 3)dx$

   $\Delta y - dy = 9\Delta x^2$

4. $\Delta y = 2x\Delta x + 5\Delta x + \Delta x^2$

   $dy = (2x + 5)dx$

   $\Delta y - dy = \Delta x^2$

5. $\Delta y = 1.001$

   $dy = 1$

   $\Delta y - dy = 0.001$

6. $\Delta y = 0.3203$

   $dy = 0.32$

   $\Delta y - dy = 0.0003$

7. $dA = 0.2 \text{ ft}^2$

   $\Delta A = 0.2001 \text{ ft}^2$

Section 3.4

1. $3\cos 3x$

2. $-11\sin 11x$

3. $8\cos 8x$

4. $-29\cos 29x$

5. $90\sin 30x$

6. $-49\sin 7x$

7. $3(2x^2 + 7x)^2(4x + 7)$

8. $39(x + 1)^{38}(1)$

9. $25(x^3 - 9x^2 + 3x + 11)^4(3x^2 - 18x + 3)$

10. $-5(3x^4 - 2x)^{-6}(12x^3 - 2)$

11. $-\dfrac{1}{2}(3x^2 + 5x)^{-\frac{3}{2}}(6x + 5)$

12. $3(\sin 10x)^2(10\cos 10x)$

13. $5(\cos 3x)^4(-3\sin 3x)$

14. $-59(\cos 2x)^{-2}(-2\sin 2x)$

15. $-66(\sin 10x)^{10}(10\cos 10x)$

16. $(9x^2 + 2)(x + 1)^3$

   $+ (3x^3 + 2x + 1)(3(x + 1)^2(1))$

17. $5(x^4 - 1)^4(4x^3)(2x^2 - x)^7$

   $+ (x^4 - 1)^5\left[7(2x^2 - x)^6(4x - 1)\right]$

18. $4x^3(x + 1)^3 + x^4\left[3(x + 1)^2(1)\right]$

19. $45(x^2 + 2x)^2(2x + 2)(2x + 1)$

   $+ \left[15(x^2 + 2x)^3(2)\right]$

20. $2(x^2 + 2)(2x)(\cos 5x)^3$

   $+ (x^2 + 2)^2\left[3(\cos 5x)^2(-5\sin 5x)\right]$

21. $\left[\dfrac{(x + 1)(2x) - (x^2)(1)}{(x + 1)^2}\right](2x^3 + 2x + 1)^3$

   $+ \left[\dfrac{x^2}{x + 1}\right]\left[3(2x^3 + 2x + 1)^2(6x^2 + 2)\right]$

22. $10x(x + 1)^3(x^2 + 2x^3 + 5x)^5$

   $+ 5x^2\left[3(x + 1)^2(1)\right](x^2 + 2x^3 + 5x)^5$

   $+ 5x^2(x + 1)^3$

   $\left[5(x^2 + 2x^3 + 5x)^4(2x + 6x^2 + 5)\right]$

23. $5\left[\dfrac{\cos 10x}{\sin^2 3x}\right]^4$

   $\left[\dfrac{(\sin^2 3x)(-10\sin 10x) - (\cos 10x)[2(\sin 3x)(3\cos 3x)]}{(\sin^2 3x)^2}\right]$

24. $\left[5(2x + 1)^4(2)\right](x^2 + 5x)^4(-3x^4 - 1)^2$

   $+ (2x + 1)^5\left[4(x^2 + 5x)^3(2x + 5)\right](-3x^4 - 1)^2$

   $+ (2x + 1)^5(x^2 + 5x)^4\left[2(-3x^4 - 1)(-12x^3)\right]$

Section 3.5

1. $y' = \dfrac{-14x - 2y}{1 + 2x}$

2. $y' = \dfrac{-x}{y}$

3. $y' = \dfrac{27x^8 + 4x - 38x - 3y^4 x^2}{7y^6 + 4y^3 x^3}$

4. $y' = \dfrac{3x^2 y^4 - 1 - 2x}{1 - 2y - 4x^3 y^3}$

5. $y' = \dfrac{1 - 3x^2(x + y) - x^3}{(x^3 - 1)}$

6. $y' = \dfrac{2\cos y}{3y^2 + 2x \sin y}$

7. $y' = \dfrac{y^2}{8(y + 5)^7 - 2xy - 4y^3}$

Section 3.6

1. $f''(x) = 56x^6$
   $f'''(x) = 336x^5$

2. $f''(x) = 24x^2$
   $f'''(x) = 48x$

3. $f''(x) = -\sin x$
   $f'''(x) = -\cos x$

4. $f''(x) = 96x^2 - 6x^{-4}$
   $f'''(x) = 192x + 24x^{-5}$

5. $f''(x) = \dfrac{2}{3}\left(x + x^2 + 5\right)^{-\frac{2}{5}}$
   $+ \left(\dfrac{1}{3} + \dfrac{2}{3}x\right)\left[-\dfrac{2}{3}\left(x + x^2 + 5\right)^{-\frac{5}{3}}(1 + 2x)\right]$

6. $f''(x) = 2\left(2x^3 - 2\right)^4$
   $+ 2(x + 1)\left[4\left(2x^3 - 2\right)^3\left(6x^2\right)\right]$
   $+ 2(x + 1)\left[24x^2\left(2x^3 - 2\right)^3\right]$
   $+ (x + 1)^2$
   $\left[48x\left(2x^3 - 2\right)^3 + 24x^2\left[3\left(2x^3 - 2\right)^2\left(6x^2\right)\right]\right]$

7. $f''(x) = -18\cos 3x$
   $f'''(x) = 54\sin 3x$

8. $f''(x) = -40(\cos 2x)(\cos 2x)$
   $\quad\quad - 40(\sin 2x)(\sin 2x)$

Chapter 4

Section 4.1

1. $(0, 2)$
2. $(0, 3)$
3. $(-2, 1991)$
4. $\left(\dfrac{4}{9}, -1.8\right)$, $(-3, 59.5)$
5. No critical points
6. No critical points
7. $\left(-\dfrac{1}{4}, -13.6\right)$
8. $\left(\dfrac{5}{3}, -6.3\right)$
9. No critical points
10. $(-0.1, 0.4375)$, $(-0.9, -3.4375)$
11. $(0,0)$
12. $(-8, 5)$, $(0, 1)$

Section 4.2

1. min: $(0, 1)$

2. max: $\left(\dfrac{1}{3}, -1.37\right)$
   min: $(2, -6)$
   inflection point: $\left(\dfrac{7}{6}, -3.7\right)$

3. max: $(1.5, 4.4)$
   inflection points: $(0, 1)$, $(1, 3)$

4. max: $\left(\sqrt{5}, \dfrac{\sqrt{5}}{5}\right)$
   min: $\left(-\sqrt{5}, -\dfrac{\sqrt{5}}{5}\right)$

inflection points: $\left(-\sqrt{15}, -\dfrac{\sqrt{15}}{10}\right)$

$$\left(\sqrt{15}, \dfrac{\sqrt{15}}{10}\right)$$

5. max / mins at endpoints: min: $(0, 2)$

max: $(4, 4)$

concave down between

Section 4.3
1. 0
2. 0
3. 1
4. 1
5. 2
6. $\dfrac{5}{7}$
7. ∞
8. ∞
9. ∞
10. ∞
11. 0
12. 2
13. ∞
14. $-\infty$
15. ∞

Section 4.4
1. min: triangle's base length = 2.19
   max: don't cut, only make circles
2. base = 47.2, height = 31.4 ft.
3. $83.30
4. radius = 2 in, height = 2 in.

Section 4.5.1
1. time to max height = 17.1 sec,
   max height = 4,692 ft.
2. round trip time = 34.2 sec.
   impact velocity = -547.2 ft/sec
   acceleration = -32 ft/sec

Section 4.5.2
1. 0.02 ft/sec.
2. 17 ft/sec.
3. 1.13 ft/sec.

Chapter 5

Section 5.2
1. x^2
2. $\dfrac{1}{2}x^2$
3. $\dfrac{1}{2}x^4$
4. $-6x^3$
5. $\dfrac{-19}{4}x^4$
6. $-48x^2$
7. $-x^{-1}$
8. $-\dfrac{1}{2}x^{-2}$
9. $\dfrac{1}{10}x^{20}$
10. $-10x^{-1}$
11. $\dfrac{1}{2}\sin 2x$
12. $-\dfrac{1}{19}\cos 19x$
13. $3x^3 + 4x^2 - 2x$
14. $-\dfrac{1}{5}x^5 - x^4 - \dfrac{10}{3}x^3 - x$
15. $\sin x + 9x^3 - \dfrac{11}{2}x^2 + 2x$
16. $-1995x$
17. $\dfrac{1}{3}(x-2)^2$
18. $\dfrac{1}{3}x^3 + \dfrac{3}{2}x^2 - 10x$
19. $\dfrac{1}{3}x^3 + 6x^2 + 20x$

20. $\dfrac{14}{9}x^{\frac{9}{7}}$

21. $10x - \dfrac{1}{2}\cos 2x - 19x^{-1}$

22. $110x - 7x^{-1} - \dfrac{51}{3}x^{-3}$

Section 5.3

1. $\dfrac{9}{2}$

2. $\dfrac{15}{4}$

3. -24

4. 26

5. 359.3

6. 56

7. 60.16

8. 15.8

9. -1

10. 0

11. -0.96

12. 1

13. $23\dfrac{2}{3}$

14. 10.1

15. 9

16. 238.3

Section 5.4

1. $\dfrac{1}{4}x^4 + C$

2. $\dfrac{1}{6}x^6 + C$

3. $\dfrac{1}{2}x^4 + C$

4. $-x^7 + C$

5. $\dfrac{1}{3}x^3 + \dfrac{5}{2}x^2 + x + C$

6. $\dfrac{1}{2}x^6 - \dfrac{1}{5}x^5 + 2x + C$

7. $\sin x + C$

8. $\dfrac{1}{3}\sin 3x - \dfrac{1}{7}\cos 7x + \dfrac{1}{3}x^3 + x + C$

9. $\dfrac{1}{4}(x+1)^4 + C$

10. $\dfrac{1}{6}(x-3)^6 + C$

11. $x - x^2 - x^{-1} + C$

12. $\dfrac{2}{3}x^{\frac{3}{2}} + \dfrac{9}{4}x^{\frac{4}{3}} + \dfrac{7}{6}x^{\frac{6}{7}} + C$

Section 5.5

1. $\dfrac{1}{32}(8x-6)^4 + C$

2. $\dfrac{1}{170}(17x+5)^{10} + C$

3. $\dfrac{1}{9}(6x+5)^{\frac{3}{2}} + C$

4. $\dfrac{3}{76}(19x+96)^{\frac{4}{3}} + C$

5. $\dfrac{1}{3}(x^2+5)^{\frac{3}{2}} + C$

6. $-\dfrac{4}{3}(18-3x^3)^{\frac{3}{2}} + C$

7. $\dfrac{1}{2}x^2 - 2x + C$

8. $-\dfrac{1}{8}(2x^2+5)^{-2} + C$

9. $-(x^2-3x+9)^{-1} + C$

10. $\dfrac{1}{5}(4x^8 - 6x^3 + 2)^5 + C$

11. $\dfrac{4}{3}(7x^5 + 4x^3 - 2x + 1)^{\frac{3}{4}} + C$

12. $-\dfrac{3}{2}\left(10 + \dfrac{1}{x}\right)^{\frac{2}{3}} + C$

13. $\dfrac{1}{18}(3\cos x - 10)^6 + C$

14. $\frac{3}{4}(\sin x)^4 + C$

15. $\frac{2}{5}\left(x^{\frac{1}{2}} + 10\right)^5 + C$

16. $-\left(\cos^2 x + 10\right)^{\frac{1}{2}} + C$

17. $\frac{1}{10}\sin(10x + 13) + C$

18. $-\frac{1}{3}\cos(3x - 2) + C$

Chapter 6

Section 6.1

1. $7\frac{1}{3}$

2. 36

3. 9.3

4. 9.3

5. 27.2

Section 6.2

1. $\frac{1}{20}y^2 + \frac{1}{5}y + C$

2. $y^2 - \frac{5}{2}Y + C$

3. $\frac{1}{2}y^2 - 2y + C$

4. $\frac{1}{3}y^3 - 5y + C$

5. $\frac{32}{3}y^3 - \frac{9}{2}y^4 - y^2 + 12y + C$

6. $8y^3 + 13$

7. 22.36

8. 42.56

Section 6.3

1. 13.73π

2. 12π

3. 8π

4. 36.27π

5. 298.2π

6. 4.72π

7. 4.72π

8. 4.12π

9. 4.12π

10. 17.1π

11. 31.8π

12. 252π

13. 268π

Chapter 7

Section 7.1

1. $\frac{1}{x}$

2. $-\frac{2}{5 - 2x}$

3. $\left(\frac{1}{8x^3 + 5x^2 + 7x + 9}\right)(24x^2 + 10x + 7)$

4. $\frac{19}{2}\left(\frac{1}{19x + 96}\right)$

5. $\frac{-\frac{1}{2}(24x^2 - 5)}{(8x^3 - 5x)}$

6. $\frac{3x}{2(3x^2 + 5)}$

7. $\frac{13}{13x + 1} + \frac{3(16x - 9)}{8x^2 - 9x}$

8. $\frac{14x - \frac{1}{3}(x + 5)^{-\frac{2}{3}}}{7x^2 - (x + 5)^{\frac{1}{3}}}$

9. $3\cot 3x$

10. $4\ln\cos 2x - 8x\tan 2x$

11. $\frac{-9x^2 - 2x - 27}{(3 - x^2)(9x + 1)}$

12. $51x^2 + \dfrac{1}{x}$

13. $\dfrac{(4\cos 4x)\ln 3x^4 - \left(\dfrac{1}{3x^4}(12x^3)\right)}{\left(\ln(3x^4)\right)^2}$

14. $2x(\sin 3x)(\ln \sin 3x)$
    $+ 3x^2\cos 3x(\ln\sin 3x) + 3x^2\cos 3x$

15. $\dfrac{1}{8}\ln|x| + C$

16. $\dfrac{1}{7}\ln|7x+5| + C$

17. $\dfrac{1}{16}\ln|8x^2+5| + C$

18. $\ln|x+3| + C$

19. $\dfrac{1}{2}(3+\ln x)^2 + C$

20. $\ln|\sin x| + C$

21. $\ln|\sec x| + C$

22. $2e^{2x}$

23. $24x^2 e^{-8x^3}$

24. $19e^{19x+96}$

25. $-19e^{-(19x+96)}$

26. $36xe^{x^2-1} + 6x^2$

27. $\dfrac{1}{3}(8+2e^{2x})(8x+e^{2x})^{-\frac{2}{3}}$

28. $6x$

29. $-2e^{2x}\sin e^{2x}$

31. $\dfrac{1}{2}e^{2x} + C$

32. $-\dfrac{1}{24}e^{-8x^3} + C$

33. $\dfrac{1}{8}e^{8x-7} + C$

34. $5\sin e^{2x} + C$

35. $-\dfrac{1}{(e^x+2)} + C$

Section 7.2

1. $\dfrac{3x^2}{(x^3+4)\ln 3}$

2. $\left(\dfrac{1}{(x^7 - x^3 + \sqrt{23x})\ln 12}\right)$
   $\left(7x^6 - 3x^2 + \dfrac{23}{2}x^{-\frac{1}{2}}\right)$

3. $\left(\dfrac{24x^2}{\sin(8x^3+5)\ln 2}\right)\left[\cos(8x^3+5)\right]$

4. $\dfrac{2}{3(2x-3)\ln 4}$

5. $16x\log_2(3x^3 - 5x^2 + 1)$
   $+(8x^2)\left[\dfrac{(9x^2-10x)}{(3x^3-5x^2+1)\ln 2}\right]$

6. $3^x \ln 3$

7. $2(x-1)19^{(x-1)^2}\ln 19$

8. $\left[(x^2+1)^{x-1}\ln(x^2+1)\right](1)$

9. $\left[(19x+96)^{3x^2}\ln(19x+96)\right](6x)$

10. $(3^{\sin x}\ln 3)\cos x$

11. $(8^{\cos^2 x}\ln 8)[2(\cos x)(-\sin x)]$

12. $\left[(\sin 3x)^{2x}\ln(\sin 3x)\right](2)$

13. $\left(\dfrac{1}{\ln 3}\right)3^x + C$

14. $\dfrac{1}{5}\left(\dfrac{1}{\ln 8}\right)8^{5x} + C$

15. $-\dfrac{1}{3}\left(\dfrac{1}{\ln 8}\right)8^{-3x}+C$

16. $\dfrac{1}{3}\left(\dfrac{1}{\ln 2}\right)2^{x^3}+C$

17. $\sin 4^x + C$

Section 7.3

1. $2\sec^2 2x$

2. $-24x^2\left[-\csc\left(-8x^3\right)\cot\left(-8x^3\right)\right]$

3. $4x\left(x^2-1\right)\sec\left[\left(x^2-1\right)^2\right]$

   $\tan\left[\left(x^2-1\right)^2\right]$

4. $-8\csc^2(8x)$

5. $\left[-\csc\left(9x^2+2x-10\right)\right]$

   $\left[\cot\left(9x^2+2x-10\right)\right](18x+2)$

6. $32x^3\tan\left(3x^4\right)+96x^7\left[\sec^2\left(3x^4\right)\right]$

7. $45\left[\cot(9x+1)\right]^4\left[-\csc^2(9x+1)\right]$

8. $\dfrac{1}{1996}\ln|\sec 1996x|+C$

9. $\dfrac{1}{8}\ln|\sin(8x+1)|+C$

10. $\dfrac{1}{3}\ln|\sec x^3+\tan x^3|+C$

11. $10\ln|\csc x-\cot x|+C$

12. $\dfrac{1}{2}\tan 2x+C$

13. $-\cot x+\tan x+C$

14. $34\sec x+C$

15. $-\csc x-\cot x+C$

Section 7.4

1. $\dfrac{3}{\sqrt{1-9x^2}}$

2. $-\dfrac{\left(36x-6x^{-3}\right)}{\sqrt{1-\left(18x^2+3x^{-2}\right)}}$

3. $\dfrac{\dfrac{1}{2}(x+1)^{-\frac{1}{2}}}{x+2}$

4. $\dfrac{1}{\sqrt{121x^2-1}}$

5. $12x^2\cos^{-1}\left(81x^2-3\right)$

   $-\dfrac{648x^4}{\sqrt{1-\left(81x^2-3\right)^2}}$

6. $\dfrac{1}{3}\tan^{-1}\left(\dfrac{x}{3}\right)+C$

7. $\dfrac{2}{\sqrt3}\tan^{-1}\left(\dfrac{\sqrt x}{\sqrt3}\right)+C$

8. $\dfrac{1}{4}\sin^{-1}4x+C$

9. $\dfrac{1}{9}\sec^{-1}\left(\dfrac{e^x}{9}\right)+C$

10. $-\sin^{-1}\left(\dfrac{\cos x}{4}\right)+C$

Section 7.5

1. $5\sinh(5x)$

2. $\left[-\operatorname{csch}^2\left(x-x^{-2}+3\right)\right]\left(1+2x^{-3}\right)$

3. $\left(\dfrac{1}{1-\left(x^3\right)^2}\right)\left(3x^2\right)$

4. $\left[\dfrac{1}{\sqrt{\left[(x+5)^3\right]^2+1}}\right]\left[3(x+5)^2(1)\right]$

5. $(\sinh x)\left(\sinh^{-1}x\right)$

   $+(\cosh x)\left(\dfrac{1}{\sqrt{x^2+1}}\right)(1)$

6. $4x^3 \sec h^{-1}(2x)$

$$+(x^4)\left[\dfrac{1}{2x\left(1-(2x)^2\right)^{\frac{1}{2}}}\right](2)$$

7. $\dfrac{1}{8}\tanh(8x)+C$

8. $-\coth x + C$

9. $\cosh^{-1}x + C$

10. $\sin^{-1}\left(\dfrac{x^{\frac{3}{2}}}{\sqrt{2}}\right)+C$

Section 7.6

1. ∞ ∴ diverges

2. $\dfrac{1}{3}$ ∴ converges

3. ∞ ∴ diverges

4. ∞ ∴ diverges

5. π ∴ converges

Section 7.7

1. ∞ ∴ diverges

2. -2 ∴ converges

3. ∞ ∴ diverges

4. ∞ ∴ diverges

5. ∞ ∴ diverges

Chapter 8

Section 8.1

1. $xe^x - e^x + C$

2. $-x^2\cos x + 2x\sin x + 2\cos x + C$

3. $\dfrac{1}{2}\left[x\cos(\ln x)+x\sin(\ln x)\right]+C$

4. $x(\ln x)^2 - 2(x\ln x - x)+C$

5. $\dfrac{1}{2}x^2\tan^{-1}x - \dfrac{1}{2}(x-\tan^{-1}x)+C$

6. $\dfrac{1}{2}\left[-\sin x\cos x + x\right]+C$

7. $x^2(x^2+1)^{\frac{1}{2}} - \dfrac{2}{3}(x^2+1)^{\frac{3}{2}}+C$

Section 8.2

1. $\dfrac{3}{8}x - \dfrac{1}{4}\sin 2x + \dfrac{1}{16}\sin 2x + C$

2. $\dfrac{1}{6}\sin^6 x - \dfrac{1}{8}\sin^8 x + C$

3. $\dfrac{1}{5}\tan^5 x - \dfrac{1}{3}\tan^3 x - x + C$

4. $\dfrac{1}{14}\sin 7x + \dfrac{1}{2}\sin x + C$

Section 8.3

1. $\ln\left|\dfrac{6}{x} - \dfrac{\sqrt{36-x^2}}{x}\right|+C$

2. $(\sqrt{x^2+9})+3\ln\left|\dfrac{\sqrt{x^2+9}}{x} - \dfrac{3}{x}\right|+C$

3. $-\left(\dfrac{x}{\sqrt{x^2-1}}\right)+C$

4. $9\sin^{-1}\dfrac{x}{3}+x\sqrt{9-x^2}+C$

5. $\dfrac{1}{250}\left[\tan^{-1}\dfrac{x}{5}+\dfrac{5x}{x^2+25}\right]+C$

Section 8.4

1. $\sin^{-1}\left(\dfrac{x-1}{\sqrt{17}}\right)+C$

2. $-\dfrac{1}{8.25}\tanh^{-1}\left(\dfrac{x-4.5}{\sqrt{8.25}}\right)+C$

3. $\dfrac{1}{\sqrt{2}}\ln\left|\dfrac{x+1}{\sqrt{17}}+\dfrac{\sqrt{(x+1)^2-17}}{\sqrt{17}}\right|+C$

Section 8.5

1. $\dfrac{1}{2}\ln|x+6| + \dfrac{3}{2}\ln|x-6| + C$

2. $-\dfrac{26}{11}\ln|x+2| + \dfrac{15}{11}\left(\dfrac{1}{x-2}\right)$

   $+ \dfrac{14}{11}\ln|x-1| + C$

3. $4\ln|x^2+2| + \dfrac{5}{\sqrt{2}}\tan^{-1}\left(\dfrac{x^2+2}{\sqrt{2}}\right)$

   $+ \dfrac{8}{x^2+2} + C$

4. $\dfrac{1}{2}x^2 - 8\left[\dfrac{1}{2}\ln|x^2+4| + \left(\dfrac{1}{x^2+4}\right)\right]$

Section 8.6

1. $\dfrac{57}{5}(x-1)^{\frac{5}{3}} + \dfrac{345}{2}(x-1)^{\frac{2}{3}} + C$

2. $\dfrac{10}{3}(\sqrt{x}+3)^3 - 45(\sqrt{x}+3)^2 + 270(\sqrt{x}+3)$

   $- 270\ln|\sqrt{x}+3| - 8(\sqrt{x}+3) + 24\ln|\sqrt{x}+3| + C$

Chapter 9

Section 9.1

1. 1
2. 0
3. $\dfrac{3}{2}$
4. $-\dfrac{1}{8}$
5. $\dfrac{1}{4}$
6. 4
7. $-\dfrac{1}{2}$
8. 0
9. $-\infty$

Section 9.2

1. 0
2. 0
3. 0
4. 1

Section 9.3

1. 6.0×10^{-6} seconds
2. 84.2 years
3. 6.58 years
4. 24.8 minutes

Chapter 10

Section 10.1

1. ∞ ∴ diverges
2. 0 ∴ converges
3. ∞ ∴ diverges
4. $\dfrac{4}{9}$ ∴ converges
5. 0 ∴ converges

Section 10.2

1. converges, sum $= 42$
2. converges, sum $= \dfrac{3}{4}$
3. converges, sum $= \dfrac{1}{14}$
4. diverges
5. diverges
6. diverges
7. diverges
8. converges
9. diverges
10. converges, sum $= \dfrac{1}{2}$
11. converges, sum $= -\dfrac{3}{10}$

12. converges

13. converges

14. diverges

Section 10.4

1. no information
2. diverges
3. converges
4. diverges
5. converges
6. diverges
7. diverges
8. converges
9. converges
10. no information
11. diverges
12. converges

Section 10.5

1. $-1 \le x < 1$
2. diverges for all x
3. converges for all x

Section 10.6

1. $1 + x + \dfrac{1}{2}x^2 + \dfrac{1}{6}x^3 + \ldots + \dfrac{1}{n!}x^n + \ldots$

2. $\cosh x = \displaystyle\sum_{k=0}^{\infty} \dfrac{1}{(2k)!} x^{2k}$

3. 0.23385

Chapter 11

Section 11.2

1. $\mathbf{r} = 26\mathbf{i} + 15\mathbf{j}$
2. $\mathbf{r} = \text{-}4.33\mathbf{i} + 2.5\mathbf{j}$
3. $\mathbf{r} = 1.73\mathbf{i} - \mathbf{j}$
4. $\mathbf{r} = \text{-}5.82\mathbf{i} - 1.45\mathbf{j}$
5. $r = 5.39,\ \theta = 68.2°$
6. $r = 67.12,\ \theta = 93.4°$

7. $r = 6.32,\ \theta = \text{-}71.6°$
8. $r = 5.39,\ \theta = 248.2°$

Section 11.3

1. $\mathbf{r} = 18\mathbf{i} + 9\mathbf{j}$
2. $\mathbf{r} = \mathbf{i} + 15\mathbf{j}$
3. $\mathbf{r} = 14\mathbf{i} - 5\mathbf{j} + 9\mathbf{k}$
4. $\mathbf{r} = -11\mathbf{i} - \mathbf{j} + 12\mathbf{k}$
5. $\mathbf{r} = 8\mathbf{i} + 9\mathbf{j} + \mathbf{k}$

6. $\mathbf{r} = -14\mathbf{i} + 7\mathbf{j} + 24\mathbf{k}$
7. $\mathbf{a} \cdot \mathbf{b} = 84,\ \mathbf{a} \times \mathbf{b} = \text{-}92\mathbf{k}$
8. $\mathbf{a} \cdot \mathbf{b} = \text{-}44,$
   $\mathbf{a} \times \mathbf{b} = -6\mathbf{i} + 9\mathbf{j} + 12\mathbf{k}$
9. $\mathbf{a} \cdot \mathbf{b} = 0,$
   $\mathbf{a} \times \mathbf{b} = 8\mathbf{i} + 8\mathbf{j}$
10. $\mathbf{a} \cdot \mathbf{b} = \text{-}460,$
    $\mathbf{a} \times \mathbf{b} = -60\mathbf{i} + 890\mathbf{j} + 380\mathbf{k}$

Chapter 12

Section 12.2

1. $\mathbf{s}'(\mathbf{t}) = 16t\mathbf{i} - 64t^7\mathbf{j}$
   $\mathbf{s}''(\mathbf{t}) = 16\mathbf{i} - 448t^6\mathbf{j}$
2. $\mathbf{s}'(\mathbf{t}) = e^t\mathbf{i} - 2(t-4)\mathbf{j}$
   $\mathbf{s}''(\mathbf{t}) = e^t\mathbf{i} - 2\mathbf{j}$

3. $\dfrac{32}{9}$

4. $e^5 - 312.67$

Section 12.3

1. speed: 0.14

   velocity: $-\dfrac{1}{9}\mathbf{i} - \dfrac{2}{25}\mathbf{j}$

   acceleration: $\dfrac{2}{27}\mathbf{i} + \dfrac{20}{625}\mathbf{j}$

2. speed: 148.4

velocity: $3\mathbf{i} + e'\mathbf{j}$

acceleration: $e'\mathbf{j}$

Chapter 13

Section 13.1

1. $f_x(x,z) = \dfrac{\partial y}{\partial x} = 36xz^3 + 2e^x z$

$f_z(x,z) = \dfrac{\partial y}{\partial z} = 2 + 54x^2z^2 + 2e^x$

$f_{xx}(x,z) = \dfrac{\partial^2 y}{\partial x^2} = 36z^3 + 2e^x z$

$f_{xz}(x,z) = \dfrac{\partial^2 y}{\partial x \partial z} = 108xz^2 + 2e^x$

$f_{zz}(x,z) = \dfrac{\partial y}{\partial z \partial z} = 108x^2 z$

$f_{zx}(x,z) = \dfrac{\partial y}{\partial z \partial x} = 108xz^2 + 2e^x$

2. $f_x(x,z) = \dfrac{\partial y}{\partial x} = 6(x + 2z)^2$

$f_z(x,z) = \dfrac{\partial y}{\partial z} = 12(x + 2z)^2$

$f_{xx}(x,z) = \dfrac{\partial^2 y}{\partial x^2} = 12(x + 2z)$

$f_{xz}(x,z) = \dfrac{\partial^2 y}{\partial x \partial z} = 24(x + 2z)$

$f_{zz}(x,z) = \dfrac{\partial y}{\partial z \partial z} = 48(x + 2z)$

$f_{zx}(x,z) = \dfrac{\partial y}{\partial z \partial x} = 24(x + 2z)$

Section 13.2

1. $f_x(r,p) = 3(2x + y) + \dfrac{3}{2}(3x + y)^{-\frac{1}{2}}$

$f_y = 3x - \dfrac{1}{2}(3x - y)^{-\frac{1}{2}}$

2. $f_x(r,p) = \left[38(e^x + y) + 2(e^x + y)(y^2 x - 1)^3\right]e^x$

$+ \left[96 + 3(e^x + y)^2(y^2 x - 1)^2\right]y^2$

$f_y = \left[38(e^x + y) + 2(e^x + y)(y^2 x - 1)^3\right]$

$+ \left[96 + 3(e^x + y)^2(y^2 x - 1)^2\right]2yx$

3. $f_x(r,p,q) = \left[x^2 y + \sin(1 + y^2)\right]$

$+ \left[(x + 2)(1 + y^2) + \left(\dfrac{1}{x^2 y}\right)\right](2xy)$

$f_y(r,p,q) = \left[(x + 2)(1 + y^2) + \left(\dfrac{1}{x^2 y}\right)\right]x^2$

$+ \left[(x + 2)x^3 y + (x + 2)\cos(1 + y^2)\right]2y$

Section 13.3

1. 34.8
2. 34.8
3. 22.9
4. -21.86

Index

X

Y

Z

Also available through Prometheus Enterprises, Inc.:

Hurricane Statistics: The New Approach To Statistics. Hurricane Statistics
is a revolutionary new book that brings clarity and insightfulness to one of the most
useful but misunderstood areas of study--statistics. For the first time ever, all of the "hard
core" statistical concepts and techniques normally taught in two semesters of college
statistics are presented simply enough for even high school students to understand while
maintaining the technical integrity required to make the book a useful reference for the
likes of research engineers. More importantly, ***Hurricane Statistics*** shows how statistics
fits into the overall scheme of science, how to use it correctly in the real world, and how
to analyze the statistical information people "in the know" use daily in an effort to
persuade people. In short, ***Hurricane Statistics*** does more than just cover the
fundamentals of statistics, it gives the reader the tools required to actually use and
analyze statistics in the real world.

TS Gradebook 2.0: A teacher-friendly Windows based software program that lets
teachers do their grades on computer. Here are just a few of its many features:

- Handles any grading system.

- Total grades calculated as individual entries are made.

- Unlimited gradebook space! No restrictions!

- Customized grade reports and printouts.

- Individual student reports.

- Report Card feature for calculating semester and year grades.

Plus, it comes with a 10-day money back guarantee!

To obtain additional information about these and other products, please contact:

Prometheus Enterprises, Inc.
P.O. Box 357
Hinckley, Ohio 44233
U.S.A.

Telephone: 1-800-393-3415

E-mail: REQ91@aol.com